HOW TO BE A
RAILWAY SIGNALMAN

Dave Walden

HOW TO BE A
RAILWAY SIGNALMAN

This Little Book version published in the UK in 2015
by Demand Media Limited

www.demand-media.co.uk

First published in the UK in 2013
by Ian Allan Publishing Ltd

Printed and bound in Europe

ISBN 978-1-910540-40-4

**RIGHT A very idyllic scene, depicting how attractive signalling on a
heritage railway can be. Semaphore signals, mechanically worked
from a nearby signall box, and steam trains going past the window.
This is Bewdley South on the Severn Valley Railway, with 34053
Sir Keith Park on loan from the Swanage Railway, heading for
Kidderminster on Friday 5 July 2013, although if the headcode discs
are to be believed, it could be a train from Victoria to Newhaven,
Waterloo to Portsmouth, or just a light engine to Eastleigh!**

Contents

Foreword

If, having visited a heritage railway and watched those possibly strange rituals concerned in the general operation of the trains, you have ever felt encouraged to become a volunteer on such a line but wondered what was involved in so doing, then, so far as the signalling of trains is concerned, this book is for you. There will always be vacancies since existing volunteers may leave due to changing personal circumstances or ill health for example, leaving gaps to be filled by new recruits.

The author is himself a long-standing volunteer signalman on a major heritage line and has set out in a very readable style the steps necessary to become a signalman, from the need to join the railway's volunteer organisation (important for insurance purposes) all the way through the departmental recruitment process and training programs (both theoretical and practical), leading eventually to a qualification to take charge of a signal box in your own right.

There is much to learn, but in reality it all comes down to 'applied common sense' and is nothing to be afraid of. Once you are qualified you will receive nothing at all for 'ordinary' time and even less for 'overtime', but you will gain an immense sense of pride and job satisfaction in a task well done, always remembering that even though the trains run slowly (up to 25mph) compared with the National Network, they are still conveying real people for an enjoyable day's outing, and in no sense will you be merely 'playing at trains'.

ABOVE After all the training and studying of the Rules and Regulations, this is what it's all about. Semaphore signals and mechanical signalling, in this case in a very rural location at Bewdley South on the Severn Valley Railway, with the signals for the line from Kidderminster on the right, and those for the Stourport siding on the left.

If this book encourages even one new recruit to join the growing number of volunteers on heritage lines then it will have achieved its aim. I commend it to you.

Russell Maiden
Signalling Manager, Severn Valley Railway
March 2013

Preface

My introduction to signalling came when a friendly British Railways signalman asked me if I would like a look in his signal box, and then began to explain how all the instruments and levers worked. I realised then that there was much more to signalling than would appear at first sight to the casual observer from the outside. In researching the material for this book, and talking to several preserved railways, I have discovered that I was not alone when I was surprised by just how involved and technical the subject can be, and just how little knowledge of signalling the average person in the street has – and how much there was a need for a book such as this.

I have tried to put in to words and pictures what is required from anyone who, having a liking for railways, has thought of becoming a signalman on one of our many heritage lines. The feedback I have had from talking to various people is that the drop-out rate amongst those who apply for such positions is considerable, and if that situation can be improved then this work will have been worthwhile. So the aim of this work is to set out, in as simple terms as possible, what anyone who has thought of volunteering his (or her!) services to be a signalman can expect to find – what is involved, and what he will be asked to do – before he reaches the stage of being allowed to work a signal box on his own, and being put in charge of train movements.

The subject is very complex, and has many variations to its theme. Of necessity, this work can only be very broad in its approach, and give what is hoped will be a satisfactory guide to anyone who wishes to take a more active part in signalling, and how it would be achieved. Each railway is different, both geographically and in its operating methods, which in turn will be dictated by the type of train service that it wishes to run, and throughout these pages I have tried to avoid much repetition, but in some cases, particularly when discussing Rules and Regulations, it is inevitable, so I hope the reader will bear with me, and understand the reasons for doing so. For anyone with thoughts of becoming a signalman on a preserved railway, after having read this book it is suggested that he has a look at the railway of his choice, from the point of view of a casual observer, and then, if still interested, approaches the organisation himself, either in person, or by mail, to see if there is likely to be a requirement for any signalling trainees in the future.

In putting this book together I have sought the views of those working on several preserved lines, and Richard Davies, Signalling Instructor on the South Devon Railway, has been most helpful, as has Adrian Lee, Chief Signalman's Inspector on the Bluebell Railway, who gave most comprehensive information on the training programme of that particular railway. Norman Hugill of the North Yorkshire Moors Railway gave some very interesting information, and the Severn Valley Railway has also been most co-operative in allowing me to photograph various installations, and items of equipment on its railway from Kidderminster to Bridgnorth, as well as allowing Russell Maiden, its Signalling Manager, to write the Foreword. Thanks are also due to the East Somerset Railway at Cranmore, and the Somerset & Dorset Heritage Railway Trust at Midsomer Norton. The Swanage Railway, in particular, Kevin Potts. To all of these organisations, and Kevin Robertson of Ian Allan Publishing, I am most grateful for the information they have supplied because it saved me from guiding the prospective reader away from the correct route to being a signalman!

In presenting this work to the reader, it would be impossible to cover every eventuality, but if you manage to read through it, and are still interested, then I urge you to pursue the subject further and seriously consider becoming a volunteer signalman on a heritage line, because it does have a high level of job satisfaction and you will be able to play a very useful role in running a preserved railway.

Dave Walden
March 2013

Bibliography

I have made reference to several works already published on signalling, when putting this book together, and if the reader is interested in learning more about the subject of signals, and their associated equipment, then the following works are recommended:

A Pictorial Record of Great Western Signalling, by Adrian Vaughan, Oxford Publishing Company
A Pictorial Record of LMS Signals, by L. G. Warburton, Oxford Publishing Company, reprinted by Noodle Books (2010)
A Pictorial Record of Southern Signals, by George Pryer, Oxford Publishing Company

An Illustrated History of Great Northern Signalling, by Michael A. Vanns, Oxford Publishing Company
BR Signalling Handbook, by Stanley Hall, Ian Allan Publishing
Two Centuries of Railway Signalling, by Geoffrey Kichenside and Alan Williams, Oxford Publishing Company

There are also several websites that can be viewed for further information, including the Severn Valley Railway (www.svr.co.uk), which has a good section on signalling, as has the Bluebell Railway at www.bluebell-railway.co.uk/bluebell/signals.

Explanation of Terms

Absolute Block
A method of signalling or operation which only permits one train over one section of line in one direction at one time. In practice this means between two signal boxes which are open.

Adverse Weather
Fog, falling snow, heavy rain, poor visibility, or darkness. (See also Fog Marking Point.)

Banner Repeater
A signal that gives a driver advance information about a signal ahead that has limited sighting.

Bell Codes
The codes that are used and sent between signal boxes on a morse tapper or a plunger, to convey messages to signal boxes on either side.

Block Instrument
An instrument used in Absolute Block signalling to indicate the state of the line between two signal boxes.

Block Post
A signal box which, when open,

breaks a long signalling section in to two shorter ones. Signal boxes are usually placed at stations or junctions, but if the section is very long they may be placed at a convenient location on plain track, to split the section and allow an increased frequency of trains.

Block Section
The portion of line between the Section Signal of one signal box, and the outermost Home Signal of the box in advance. (See also Token Section.)

Block Shelf
A long piece of wood hung horizontally from the ceiling containing signalling instruments and indicators.

Block Switch
A switch that allows a signal box to be closed, so that adjacent signal boxes can go in to through communication with each other.

Bullhead Rail
A pattern of rail that is similar in profile at the top and bottom.

Cant
The amount of super-elevation given to the track on curves to allow faster speeds and a smoother ride.

Catch Points
Sprung points, not worked from a signal box, that are designed to derail vehicles running away in the wrong direction. Usually placed on gradients of double line railways to prevent vehicles running away downhill. (See also Trap Points.)

Clearing Point
The distance ahead of the outermost Home Signal, usually 440yd, to which the line must be clear before a train can be accepted from the signal box in the rear. Normally used in Absolute Block Working on double track, but can sometimes be used in working single lines.

Detection
A means of ensuring that facing points are correctly closed and locked before a signal is cleared.

Detonator
A small round explosive device

EXPLANATION OF TERMS

designed to be clipped to the top of the rail in emergency situations, which is set off as a Warning signal when a train runs over it. A detonator can cause serious injury when it explodes, to anyone standing closer than 30yd.

Disabled Train
A train which cannot move under its own power, and may need assistance.

Duty Officer
The person appointed at the time to be in charge of train movements on the railway.

Electric Key, Staff or Token
A physical object giving the driver authority to be on a section of line between two signal boxes on a single line, having the name of the two signal boxes to which it applies engraved on it.

Examination of Line
A procedure for ensuring that a portion of line is safe for the passage of trains.

Facing Point Lock
Equipment that ensures that points

cannot be moved irregularly or underneath a train.

Facing Points
Points which alter the direction of travel of trains. More accurately known as Facing Turnouts

Fail-Safe
All signalling equipment is designed so that if it were to fail, it would put the signals to danger. If, for example, a signalling wire was to break, the signal would automatically return to the danger position.

Fixed Distant
A Distant Signal at a permanent location, with its semaphore arm in a fixed position displaying a permanent caution indication. Usually used at junctions for the diverging route, or on the approach to terminal stations.

Fixed Signal
A permanent signal, as opposed to a hand signal or flag, in a fixed location, and identified by its type and location.

Flat-Bottom Rail
A pattern of rail that is wider at the

foot than the top, and needs fewer fixings.

Fog Marking Point
The specific point, usually a signal listed in the Signal Box Footnotes, to which the signalman must be able to see clearly, without having to introduce Fog Working. If no specific point is listed, then the signalman must be able to see clearly for a distance of 200yd. (See also Adverse Weather.)

Footnotes
A list of Special Instructions and conditions applicable to that particular signal box.

Fouling Bar
Mechanical equipment along the inside of a running rail to ensure vehicles are clear of any points to be moved.

Four-foot
The space between the running rails.

Frame
The equipment in a signal box containing all the levers for working the signals and points.

Gradient
The vertical rise and fall of the track.

Gradient Profile

A graph kept in a signal box to show the rise and fall of the railway, as well as mile post indications, and other important structures, between signal boxes.

Ground Frame

A small Lever Frame, or a number of switches, giving local access to points or sidings, but only with the permission of the controlling signal box. Access to a ground frame on a single line is usually achieved by inserting the Key/Staff/Token into a lock to release the levers. They can be in the open air, or in a hut, or on platforms. Some signal boxes are able to be worked as ground frames when the box is switched out.

Hand Signals

A signal given by hand, or with a flag or lamp, by a hand signalman from a location that may vary according to the circumstances at the time.

Home Signal

The first stop (red) signal the driver sees on the approach to a signal box. May also be called an Outer Home.

Illuminated Diagram

A diagram of the track controlled by the signal box, containing light bulbs which illuminate to show the position of the trains.

Interlocking

The equipment used to prevent the unsafe setting of routes and release of signals for the passage of trains and also to prevent the establishing of conflicting moves.

King Lever

A lever in a signal box which releases other levers and allows the signal box to work with signal boxes on either side. (*See* page 114.)

Lever Collar

A round disc with a hole in the centre, which is placed over a lever as a reminder appliance, to prevent it being moved. Designs may differ between different railway companies. (See also Reminder Appliances.)

Obstruction

Anything placed on or near the line which would prevent the safe passage of a train.

One Engine in Steam

Sometimes known as 'One Train Working'. A method of working a single line where the line is only occupied by one train, or two or more engines coupled together which are then to be treated as one train.

Pilotman

A person appointed to conduct trains over a single line, during the failure of equipment, wearing a coloured armband indicating his role, but working under the instructions of the signalman.

Protection

Action taken to stop the potential of another train running in to a failed or disabled train, or other obstruction on the line. Usually three detonators a specific distance from the obstruction. The distances may vary according to the situation, and local railway company rules.

Reminder Appliances

A device used to remind the

EXPLANATION OF TERMS

signalman of a situation requiring caution, when signalling trains. It may be a lever collar, or a flap on a Block Instrument, or some other object. (See also Lever Collar.)

Right-Road Running

Trains running in the right direction on the right line.

Roster

A sheet produced by the Roster Clerk, showing when and where signalmen are required to report for duty.

Runaway Train

A train which the driver is unable to bring under control; usually a loose-coupled freight train which has limited brake power.

Running Dummy

A shunting signal which has to show a proceed indication, before a running signal can be cleared, to allow a driver to proceed along the main line.

Running Signals

A signal such as a distant or a stop arm, sighted, observed and obeyed on the run.

S&T Department

Signalling and Telegraph Department. The department responsible for the maintenance of all the signalling equipment on the railway, and to whom all faults should be reported.

Scotches

Items placed next to the wheels, to prevent stabled vehicles moving, or placed between point blades to prevent them being moved.

Section Signal

The most advanced signal ahead of a signal box, controlling the entrance to the section of line ahead to the next signal box. The Section Signal is usually the Starting Signal, but some signal boxes have advanced starting, or even outer advanced Starting Signals, according to the track layout, but the Section Signal is always the most advanced signal before the driver enters the section ahead.

Section Times

The length of time it takes a train to travel from adjacent signal boxes. It may vary according to the type of train.

Sequential Locking

A method of interlocking between levers to ensure that they are worked in the correct sequence.

Shunting Signals

Signals used in shunting, where the train has been brought to a stand and movements are being made slowly. When cleared, shunting signals only authorise the driver to proceed as far as the line is clear.

Single Line 'Token'

The generic term for a staff, key, tablet or token, authorising the driver to occupy a section of line.

Six-Foot

The space between one line of rails and another, on a double-track section of line.

Slotting

Equipment used where the Distant Signal of an adjacent signal box is on the same post as the Section Signal of the next box, to ensure that the distant arm will not show clear unless the stop arm also shows clear.

Special Traffic Notices

Notices published on the railway to

advise of train workings other than those in the Working Timetable.

Staff and Ticket

A method of working trains over a single line, where the driver *must* be shown the Staff itself, and is then given a ticket authorising him to be on the line. Usually used where there is a succession of trains to travel over a single line in the same direction, the last one taking the Staff, to permit trains to travel in the opposite direction from the other end.

Station Limits

The portion of line between the outermost signal on the approach side of a signal box, to the most advanced signal ahead of the signal box, in the same direction, worked from that signal box. The limits will be different in the opposite direction, because of the different location of the signals.

Subsidiary Signals

A miniature semaphore arm, permitting movement, when cleared, in to an occupied line, or ahead of a main arm for shunting purposes.

Time Interval Working

A method of working, used as a last resort, when all communication with other signal boxes has been lost, to work trains over a line, based on the time the last train departed and the time it should take to reach the next signal box.

Token Section

The portion of line between the Section Signal of one signal box and the outermost Home Signal of the signal box in advance, in the direction of travel, on a line of railway controlled by a Staff, Key, or Token. (See also Block Section.)

Track Circuit

A method of detecting the position of trains by running a small electric current through the rails.

Track Circuit Clip

A length of wire with clips either end, which is clipped to the rails as a safety device to protect a train or obstruction.

Trailing Points

Points which converge the travel of trains.

Train Register

The book in a signal box which the signalman is required to complete giving details of all train movements, times, and any unusual occurrences.

Trap Points

A switch blade at the exit from a siding, normally open and worked from a signal box, designed to derail unauthorised movements on to a running line. (See also Catch Points.)

Wire Adjuster

A device used to adjust the tension of signal wires during hot or cold weather, to ensure that the signals display the correct indication.

Working Timetable

An internal railway publication giving times of all scheduled movements on the railway.

Wrong Direction Running

Used on double track, where trains travel over lines in the wrong direction. Only to be used where authorised by the Duty Officer, or laid down in the Footnotes.

Chapter 1

First Requirements

Before we Begin

All heritage railways are constantly appealing for more volunteers, because there is always more work to be done than there are people to do it, and so anyone wishing to make their services available, and offer assistance, will most likely be made very welcome. This will entail becoming a member of the organisation, which may be divided into several broad categories, one being just membership of the railway with an annual subscription, and the other being a member of the volunteer working group and attending regularly to carry out certain duties, as required, according to which department of the railway the volunteer wishes to work for. Volunteers are classed as employees of the railway, but unpaid, and as such are bound by insurance and Health and Safety requirements, as well as any company rules.

Medical

Different railways will have different approaches to how they handle their volunteers, but almost certainly anyone wishing to get involved with moving trains, and wishing to be a signalman, will be required to have a medical, either carried out by the railway's doctor or by the volunteer's own doctor, if he is recognised by the railway, to establish that he is physically fit enough to carry out the tasks involved. These medicals involve eyesight and colour-blindness tests, as well as checking blood pressure and medical history. They are then repeated, initially every five years for the younger members, getting more frequent as the years progress, until after the age of 60 it may become an annual requirement. After 70 the signalman may only be allowed to continue working in a signal box by special permission of the doctor and Signalmen's Inspector.

Starting Out

Some railways ask their volunteers to start at the bottom, as a porter, so that they can be trained and made aware of the issues involved in safety critical work and working with moving trains. Having

BELOW First find yourself a railway, become a member and the rest is up to you! Some heritage lines operate a gradual progression based on experience gained often on the platform, before moving on to the operating grades. Running and indeed maintaining a service on a preserved line, regardless of the type of motive power used or level of service, is very different once you are involved.

established that they are competent in such matters, the railway then allows them to progress on to more responsible positions.

Training Course

There is always a requirement for more signalmen, because even if the number is up to strength people move on to other walks of life, or retire through age, so there is a constant turnover of staff. It is quite probable that it will be announced internally when the railway is running a signalling training programme, and volunteers will be invited to apply for positions on the course. This may mean an interview with the Training Instructor, or a Signalmen's Inspector, to assess the applicant's suitability for the position, because many people who apply do not realise exactly what is involved, especially in the knowledge of the Rules and Regulations and the physical act of pulling heavy levers.

Rules have been in existence ever since railways were built, for the guidance of their employees, and Regulations were drawn up in order to have a uniform system of running trains. These became even more essential as traffic increased and needed controlling. They also became modified as a result of incidents on the line, and have been developed over more than 150 years.

Once the numbers on a course have been decided, it is quite probable that the applicants will be taken in

FIRST REQUIREMENTS

ABOVE To an individual faced with 'learning the business' for the first time, signalling especially can be a daunting process. Take solace in the fact that others have had to learn before you, some are slow learners others may be faster, but persevere and it will all pay off. The hallmark of a good trainer is also one who can recognise the strengths and weaknesses of his class, tailoring the instruction to suit. Signalling is not learned in a day, a week or probably a month. But speaking from experience it is one of the most enjoyable aspects of railway operating and once understood will probably be retained for a lifetime. This is the training room on the Severn Valley Railway.

ABOVE & RIGHT We have come a long way from 'Go away and learn this book...'. Nowadays most railways operate a specific training programme for signalman/guards etc. In the example seen here instruments of the type to be used are available for the trainee to practice upon, including a Token machine (right) and (above, left to right) an early form of track circuit indicator, '1947 Block Instrument', Block bell. All of these have their peculiarities, but each will be learned.

to a signal box to show them what is entailed in working a Lever Frame and regulating trains, and it would be fair to say that, having discovered the realities of signalling, the drop-out rate can be quite high at all stages of the course.

Having been shown what is involved, the trainees then study the Rules and Regulations, and this is where it can become very hard work mentally, because there is a lot to learn. The best way is to take it one step at a time, if possible with a little practical tuition at the same time to keep the interest alive. Trainees will need to study the Rules applicable to

signals and signalling, and the Regulations, according to the method or methods of signalling employed on the particular railway. At the end there is likely to be a written examination on all aspects covered in the training programme. Depending on the level of success achieved, the candidate will then be allowed to train on a signal box frame with a qualified signalman. Should trainees not reach the standard required, all is not lost, however, as they will be given the chance for further training, and the opportunity either to take the written exam again, or have an oral interview with a Signalmen's Inspector.

Learning a Signal Box and Frame

Once a prospective signalman has passed his Rules and Regulations examination, he will then be put in a signal box to learn the frame. It will probably be the lowest of the grades of signal box on the line, so that he can start at the bottom and progress upwards through the grades, as the years of experience allow.

Learning a frame involves getting to know what each lever does, in which sequence the levers have to be pulled, and acquiring the knack of pulling each one. They can all be different and some of them quite heavy, depending on what they are required to do.

Learning a signal box means getting to know all the individual circumstances which apply at that particular location, how long it takes a train to travel to and from the signal boxes on either side, and any Special Instructions laid down for the working of the box, known as the Footnotes.

LEFT Having learned the frame the signalman here is shown in the process of passing out: being examined by the Signalling Manager. This is Bewdley North on the Severn Valley, certainly not a box for the novice but one which is a joy to work. As Signalling Manager (the role used to be known as District or Signalling Inspector) Russell Maiden will watch every aspect of the trainee's work: knowing which levers to pull, in the correct order and the effort required. The same will apply to the various instruments and indicators. In this view the signalman is operating one of the four block bells in this box.

FIRST REQUIREMENTS

Passing Out

Having had several training turns in a box, the trainee will be then subject to a box examination, which entails working the signal box in the presence of a Signalmen's Inspector, demonstrating that he can cope with specific situations and knows what to do should any emergency or out of course situation arise. If all goes well, he will be passed out by the inspector, and can at last 'fly solo' and work that signal box on his own. He will need to go through a similar procedure for every signal box in which he wishes to work.

LEFT The signalman will also be checked on his work outside. It may appear here as friendly banter with the crew – as indeed will often be the case – but this is also the time when an exchange of necessary information might be given between both parties.

FIRST REQUIREMENTS

Rostering

Different railways may have different procedures, but having become a fully qualified signalman he may be required to work a certain number of turns per year to maintain his certificate of competence. To achieve this, some railways have a Roster Clerk who will send out availability sheets, asking which dates signalmen are prepared to work. These sheets are then returned to the clerk, who completes a roster sheet, which is sent to all signalmen, showing on which dates they are required to work. Questions will be asked, and disciplinary action may follow, if a person does not turn up for duty on the appointed day without good reason.

Moving On

All being well, if a newly-qualified signalman performs his duties correctly, after a certain length of time, probably a year, and a minimum number of turns, he will be able to apply for training to work other signal boxes of a more complicated nature. Eventually he will know every signal box on the line and in effect be a relief signalman, able to instruct other trainees in turn.

Roster - Issue 1 — **March 2013**

DATE	SERVICE	KR	BS	BN	AY	HY	HL	BH
Sat 2	A	R.Edwards	P.Sharpe	R.Phelps	K.Lambert	T.Chamberlain	11.35-2/40	
Sun 3	A	P.Davenport	T.Furber	C.Boxall	M.Faarup	R.Goundry	S.Ward	J.Pilborough
Sun 3	Training			xxxxx	N.Wright	R.Smith		
Sat 9	A				9.55-4/10		xxxxx	R.Palmer
Sat 9	Training		P.Davenport	D.Clarke	J.Mort	T.Williams	11.55-2/40	
Sun 10	A		J.Green	A.Williams	xxxxx		I.Binner	S.Tull
Sun 10	Refresher	M.Ratcliffe		P.Barrett	M.Breeze	B.Alston	xxxxx	H.Bowling
Sun 10	Training						T.Chamberlain	D.Clarke
Sat 16	A=Charter		C.Boxall	xxxxx	xxxxx			M.Degg
Sat 16	Charter	T.Thomas	R.Maiden	C.Hall	J.Taylor	Dan.Evans	M.Crumpton	xxxxx
Sun 17	A	M.Degg		L.Mortimer	xxxxx	xxxxx	R.Dunn	J.Green
Sun 17	Training	A.Hitchmann	P.Sharpe	P.Davenport	H.Bowling	R.Goundry	S.Ward	J.Mort
				L.Mortimer				P.Ridley
Tue 19	Charter=FE	8.35-6/40	8.35-6/40		9.30-6/00	9.30-6/00	9.30-6/15	8.35-6/15
Tue 19	Training	R.Heap	P.Hale	I.Duncan	xxxxx	D.J.Evans	xxxxx	S.Corkery
			B.Holmes		xxxxx		xxxxx	H.Bowling
Wed 20	Charter	10.05-5/30	9.30-5/20	9.30-5/55				9.30-6/15
Wed 20	Training	L.Greenwood	P.Hale	R.Phelps	xxxxx	3.30-8/15	xxxxx	B.Craven
				xxxxx	xxxxx	J.Kane	xxxxx	
Thu 21	Charter=FE	9.30-5/30	9.30-5/30	7.40-5/00			9.30-6/15	xxxxx
Thu 21	Training	R.Heap	B.Curtis	R.Phelps	xxxxx		T.Crabtree	T.Chamberlain
				xxxxx	xxxxx		xxxxx	9.30-6/15
Fri 22	Gala Early	7.30-3/45	7.30-3/45	8.40-3/20	9.50-9/20	8.10-3/30	9.20-3/35	J.Watt
Fri 22	Gala Early	R.Edwards	B.Curtis	P.Davenport	M.Faarup	L.Mortimer	M.Jefferies	7.40-3/20
		D.Clarke	I.Duncan	3/25-10/15		9/20-10/11	3/35-9/50	J.Green
Sat 23	Gala Early	7.30-3/45	3.30-3/45	8.40-3/25	xxxxx	9.50-3/20	N.Kimberlin	S.Tull
Sat 23	Gala Late	C.Hall	A.Ribbands	P.Sharpe	9.50-3/25	3.30-3/30	7.30-3/20	C.Boxall
		3/45-10/15	3/35-10/05	3/25-10/15	B.Craven	Dan.Evans	T.Chamberlain	3.20-10/15
Sun 24	Gala	T.Thomas	D.Clarke	J.Phillips		9.20-10/11	3/35-9/50	P.Barrett
Mon 25	Travelling Sman	A.Hitchman	J.Harris	M.Ratcliffe	B.Holmes	N.Wright	A.Williams	K.Hammond
Tue 26	Travelling Sman							C.Hall
Wed 27	Travelling Sman							J.Harris
Thu 28	Travelling Sman		J.Harris					
Fri 29	B=FE	M.Ratcliffe	L.Greenwood	I.Duncan	C.Hall	D.J.Evans	Dan.Evans	T.Crabtree
Fri 29	Training		C.Boxall	J.Watt	B.Alston		xxxxx	
Sat 30	C	A.Hitchman	B.Curtis	N.Hopkins	J.Mort	T.Williams	D.Pugh	N.Kimberlin
Sat 30	Training	C.Owen			xxxxx	M.Faarup		R.Walker
Sun 31	C	T.Thomas	B.Curtis	P.Barrett	J.Taylor	Dan.Evans	I.Binner	B.Holmes
Sun 31	Training			xxxxx	xxxxx		xxxxx	J.Mort

ROSTER CLERK (ISSUE 1 - 10/02/13)

ABOVE All the hard work has paid off and your name appears on the roster sheet for the first time. Don't forget to apply for double-time on Sundays!

Chapter 2

Methods of Signalling

Some of our heritage railways take over existing track, but others start from the very beginning and have to build a completely new railway. Either way, the day eventually comes when the operators want to start running trains, but before that the Government Inspectors must come along to inspect the line to make sure it is safe for the carriage of passengers. This means that the railway has to meet certain safety standards before it can operate as a passenger-carrying line.

One of those requirements is that all facing points must be equipped with locks, to prevent them moving under a train, and that there should be no more facing points than is necessary. A point is said to be facing when it can change the direction in which a train is travelling. There should also be detection, between the points and signals, so that the signals cannot be put in the clear position for a route which is not set and locked, and there should also be interlocking between the points so that conflicting movements cannot be set up and likewise cannot be signalled. Once the railway has met with the approval of the Inspector, it is then permissible to run trains, but to achieve this it is necessary to establish what is called 'Block Working'.

Block Working

When the railways were built, a man was positioned at the base of a signal to turn it to danger for a specified length of time after a train had gone past him, then after a further length of time turned it to caution, and then after another lapse of time turned it to clear. This system is known as 'Time Interval Working' and is still practised today in cases of extreme emergency. The system was fraught with danger because, as the telegraph had not yet been invented, there was no means of communication with the man in advance, to know whether or not the train had reached him, so if the first train had broken down or stopped for any reason before arrival at his signal, the second one was sent forward with a likelihood of collision.

The coming of the telegraph changed all that, so the signalmen could now keep in touch with one another and knew where the trains were. So the railway became divided up into sections, or 'Blocks', between signal boxes that were open, and the principle was established of having only one train in one section, or Block, at any one time. This principle still holds today, and is indeed an Inspectorate requirement.

The definition of a section, or Block, is that portion of line from the most advanced signal of one signal box to the outermost signal in rear of the next signal box that is open. Where signal boxes are close together, these sections can be very short, but where traffic is very light, or signal boxes are switched out, some sections may be 20 miles long, or more.

To understand how the block system works on single line, it is best to describe it using three signal boxes – A, B, and C. When the man at A has a train that he wishes to despatch to B, he asks the permission of B to send it forward. This is done by using a morse tapper, and tapping out the correct bell-code for the type of train that A wishes to send, a stopping passenger train, for example, having a bell-code of 3pause1. The man at B, if the circumstances are in order at his signal box, will accept the train by answering the bell code from A, which will give A a release on his single line instrument, and allow him to withdraw a Token, which is the driver's authority to be on the line. As the train leaves A, the man there tells B it is leaving by sending another message on his morse tapper – 'train entering section', usually two beats –

which B acknowledges. While the train is between A and B, B then offers the train to C, who accepts the train. At B the Tokens (A-B, B-C) are exchanged. When the train is heading towards C, B tells C it is on its way, and then places the Token brought by the train from A into his instrument, and taps out the 'train out of section' bell-code – 2pause1 – to tell the man at A that the train he dispatched has arrived. The section is now clear, ready for the whole process to be repeated if A has another train, or to be gone through in the opposite direction if B wishes to send one to A. All bell signals, with a few exceptions, should be acknowledged by repetition, and must not be considered as understood, unless correctly repeated.

For a signalman to be able to accept a train, when offered to him, certain conditions have to be met before he can allow this to happen. Presuming that the signal box is on a single line, and not somewhere that trains can cross, the first requirement is that no train must have been accepted in the opposite direction. The second requirement is that the line must be clear to what is called a 'Clearing Point,' which is usually 440yd in advance of the signalman's outermost signal in rear, and all the points within that distance are set and locked for the passage of the train being offered. In certain instances the distance of 440yd may be modified according to the local circumstances eg where passenger trains are not be permitted to cross, but where there are goods loops or sidings and it is permissible to cross a passenger train with a freight train.

Where passenger trains are permitted to cross, the point at which the line has to be clear before a train can be accepted is generally the signal at the advance end of the loop line, on which the train has to run. Signals in rear of the loop are left at danger until the speed of the train is sufficiently reduced, to enable it to safely stop at the loop Starting Signal. If two trains are arriving together, they have to be brought in to the station one at a time. Not until the first one has been brought to a stand in the loop, can the second one enter the platform, or vice versa. At some crossing places there are 'Trap Points' at the end of each loop line to prevent a collision with an oncoming train, should the train entering the loop over-run the signal at the far end. In these circumstances, it may be in order for the signalman to clear the signals for the train entering the station, before it actually arrives, in order to prevent bringing it at or nearly to a stand at the signals approaching the loop.

Methods of Single Line Control

There are several methods of controlling trains on single lines, all of which have been developed as railways expanded and flexibility of operation became necessary. These systems vary from the minimum requirements, where only one train is operated on a line with a Staff, to Staff and Ticket Working, and Tablet or Token Working. Each will be considered in turn,

starting with the most basic form and then progressing to the most commonly used.

One Engine in Steam

The simplest method of operating a railway is to only have one train in operation, and it can then run up and down the line as much as it likes, because the whole railway is in effect one section, or Block. This system is known as 'One Engine in Steam', or 'One Train Working'. This method requires minimal, if any, signalling, but does need some form of authority for the driver to proceed along the railway. It usually takes the form of a

ABOVE An example of a train Staff used where One Engine in Steam working is in operation. This is the Staff used when the Severn Valley Railway only ran as far as Foley Park, and was the driver's authority to be on the line. The left-hand end was put in to an instrument in the signal box to release the signals, and was then removed and given to the driver. The right-hand end was used for inserting in to a Ground Frame at Foley Park to close the Trap Points if the train was to proceed on to the National network.

Staff, generally of wood, with the name of the section to which it applies carved on it, and a metal Key on the end of it. The Key is inserted into an instrument, and turned through 90 degrees, to obtain a release, if required, for any signals. The Staff is given to the driver, and the fact that he carries it with him prevents another train being on the same line. While he is on the line, no other train can enter that same section of track.

However, this method of working is very restrictive because before another train can be on the line, the Staff has to be returned to the instrument from which it came, which means that the train that took it in the first place has to bring it back. This is perfectly acceptable, provided everything runs to the timetable, but does not allow for late running, or train failures, so a more flexible method of operation is desirable.

Staff and Ticket

The next stage is a system which permits two or more trains to be on the railway at the same time, but this requires some way of keeping them apart, usually using signals, and a method of ensuring that there is only one train on any particular section of line at a time. There are several ways of achieving this, but nearly all involve having a Staff, a Tablet or a Token. With 'Staff and Ticket', the principle is the same as One Engine in Steam, but the signal box in advance will have a similar instrument to the one in the box where the train starts its journey. This allows the Staff to be put in the advance box instrument, in order to allow a train to enter the section from the far end once the first train has arrived. Should a second train need to follow the first one through the section, the driver of the first train is given a written ticket by the signalman, who retains the Staff, and the ticket is the necessary authority for the driver to proceed once the signal is cleared and the guard has given the 'Right Away'. This is known as 'Staff and Ticket Working', and the process is repeated for any subsequent trains which may be required to travel in the same direction until, with the last one, the driver takes the Staff (not a ticket) in order to allow trains to proceed in the opposite direction, once he has reached the signal box in advance.

This method gives more flexibility of operation as it permits trains to enter the section from either end, but is still not ideal as it depends on everything running to the timetable, and does not give any scope for additional safety features to be incorporated in to the equipment in the signal box, should that become necessary.

Electric Key, Staff or Tablet

When signalling engineers realised the limitations of Staff and Ticket Working, an even more flexible system was developed, not only permitting one train to be in a section at a time but, if no train was in the section, to allow a train to enter the section from either end. This system took several forms: a Key to the section, a Staff,

or a Tablet. All worked on the same basic principle of having two similar machines, one in each signal box at the end of the section. Each was electrically locked, requiring the co-operation of both signalmen to release the Key, Staff or Tablet, and once one was removed from an instrument, another one could not be taken out until the first one had been replaced, although it could be replaced in either instrument. The Key system is as it says, a Key to the section of line, and can take several forms. Some look like a Key similar to that used in an ordinary mortise lock, others, particularly on the Great Western, are of a special design, about two inches square, with a collar on the end that has a notch in it, so it can only be removed or replaced in its correct instrument. The Staff looks like a rod about 23in in length, with rings along it in certain places, again so it can only be used in specific machines. The Tablet has the appearance of a large

ABOVE A single-line railway may be operated by one of several means: One Engine in Steam, Train Staff, Staff and Ticket, Electric Token/Tablet or Acceptance Lever (see page 89). Of these by far the most common is the electric Key Token. Four colours – Red, Yellow, Blue and Green – are used, each having a different notch in the machine end which ensures they may only be used in an instrument with a corresponding cut-out. Note the section of line to which the Token applies is also engraved on the item. In recent times most Key Tokens have been made from aluminium although some of the die cast examples may still be used. On the shelf below are examples of brass Tablets, again with the section of line engraved and again different notches which permitted only the correct Tablet to be placed in the correct machine.

biscuit, with a hole in the centre, and a notch cut in the edge, to ensure that, like the others, it can only be placed in its correct machine. All of them will have the name of the section of line to which they apply

engraved on them, and all of the instruments will have a magazine, containing several Keys, Staffs or Tablets, usually 20 per section, but not always equally balanced between instruments, so permitting trains to follow one another in the same direction if required, or to enter the section from either end should there be no train in the section.

These systems gave the necessary security required to run traffic over single lines and prevent more than one train being in a section at a time, but also gave the flexibility to allow for out-of-course running and train failures.

In some signal boxes there was more than one type of system. For example, the box could use the Key method towards a box on one side and a Staff method to a signal box on the other side, or a signal box may have controlled a junction where more than one single line met and a different method was employed on each line.

Where it was possible to switch signal boxes out, in times of infrequent traffic, some signal boxes had

ABOVE Pairs of Tyers Electric Token Machines. The sharp eyed reader will notice a minor variation in the reminder appliance on the left, the pair on the right being of the older GWR type, the more modern variant having a plastic reminder knob. Unused Tokens are stored in the slots. An identical instrument able to give out/receive a Token will be located at the other end of the single line section. Only one Token may be obtained for the section at a time – one blue Token or one yellow Token. The experienced signalman (apologies for the apparent gender reference, but mechanical signalling is to many 'traditional signalling' and so the traditional name is used in preference to the term 'Signaller') will also rotate which Token is used, meaning he will not use the same one all day. A Token is withdrawn by lifting it up to the centre slot from where, with the consent of the signalman at the opposite end of the section, it may be withdrawn to be given to the driver of the train. To the right of the centre slot is the bell-push, on which the bell codes – the manner of communicating with the signalman at the other end of the section – are made.

more than one instrument, of the same type, in order to work trains over the longer sections between boxes. In these cases, the notch in the Key or Tablet, or rings on the Staff, being in a different position, would prevent the wrong Key being inserted in the wrong instrument, and so avoid it being used for the incorrect section of line.

Absolute Block

Absolute Block is the term used to describe the method of signalling on sections of track that are worked as double line. 'Absolute' refers to the absolute prohibition of there being no more than one train in a section at one time. In this case, there is, of course, no need for single line instruments. These are replaced by what are called Block Instruments, and can be of various designs, according to the manufacturer, but most will show three indications – Normal, Line Clear, and Train on Line – depending on the state of the traffic. To understand how the procedure works, it is necessary to refer back to A, B, and C signal boxes.

When there are no trains on the line, the indicator in the Block Instrument will show Normal, and the line should be considered as blocked. If the signalman in box A wishes to send a train to box B, he will offer the train to B on the bell tapper, and B, if he is prepared to accept the train, will turn the indicator on his instrument to Line Clear, which will repeat on the instrument in box A, and also give an electrical release to the Section Signal in box

A. As the train passes from box A towards B, A sends the Train Entering Section bell signal to B, who acknowledges it, and then turns his instrument to show Train on Line. This immediately destroys the electrical release that was given to the Starting Signal at A, so that the man there cannot pull his Section Signal again until he has received another Line Clear release from B. While the train is travelling towards B, the man there asks permission to send the train to C, who gives B a Line Clear, and as the train passes B, B sends Train Entering Section to C, and then, having seen the train pass, complete with tail lamp, sends Train Out of Section to A, and restores the instrument indicator to Normal, ready to go through the whole process again, if there is another train following. Should a signalman forget to send the Train Entering Section bell signal, or the signalman in advance forget to acknowledge it, or not turn his instrument to Train on Line, when the train approaches the outermost signal in rear of the box in advance there may be some form of control, activated by a train operating a track circuit, which automatically puts the Block Instrument indicator to Train on Line to prevent a following train entering the section while the first one is still there, and also to prevent the man at B mistakenly sending Train Out of Section to the man at A while he still has a train in section.

Most later types of Block Instrument have upper and lower indications to show the state of the line. The lower indications are controlled by the signalman himself and consist of an arrow pointing vertically downwards for

METHODS OF SIGNALLING

RIGHT A GWR '1947' Block Instrument. Different examples exist made in different woods and were mostly constructed at the former GWR Signal Works at Reading. This type of instrument is used for double line working – an up and down line between two signal boxes. The length of line between these two boxes is called, would you believe, the 'section'. (Told you it was sometimes easy!) Again, a similar instrument would be provided at the opposite end. A signalman is always in charge of trains approaching him; therefore the signalman 'in rear' (the direction from which the train is arriving) will ask 'Is Line Clear?' and if the train can be accepted the signalman here will turn the commutator – the black knob – at which point the needle in the lower window will move to 'Line Clear'. When the train is actually on its way between the two signal boxes, the lever will be turned in the opposite direction to indicate Train on Line. These indications are repeated in the instruments at the opposite end. (Fear not – this is double-line working; we have to be fully conversant with single line first, but it does show where you could be heading ...)

Normal, to the left for Line Clear, and to the right for Train on Line, with an operating handle beneath the arrow, which is turned to the position desired to be shown by the arrow. The upper indicator is a slave unit, which repeats the indication given by the signalman in the box in advance. The operating handle may have flaps alongside it, one on each side, which are turned through 90 degrees to prevent movement of the handle should the signalman's Clearing Point be fouled, or should there be a train stood at his signal in rear.

Some earlier types of instrument did not have a slave unit, or an arrow, but a small window with a flap behind it that rotated as the signalman worked the instrument, either Line Clear (coloured white) or Train on Line (coloured red) appearing in the window as required. When a Normal indication was required the flap gave no definite indication in the window but hung between Line Clear and Train on Line. These were known as 'Pegging' instruments and there was always another 'Non-Pegging' instrument, generally alongside it, of similar design which repeated the indications from the box in advance. If this form of signalling was worked on either side of a signal box there may have been two types of instrument, two types of block bell, both with different tones, so the signalman could distinguish which bell had rung and, consequently, which one had to be answered.

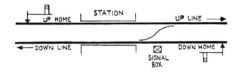

HOME SIGNALS.

BLOCK SYSTEM

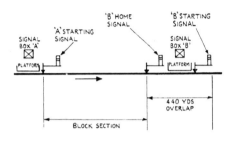

ABOVE The basic principles of Block Working, showing the positioning of signals and their terminology.

RIGHT Positioning for 'Stop' signals in their respective guise as Home', 'Starting' and 'Advanced Starting' signals.

STARTING SIGNALS.

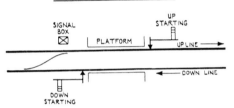

DISTANT, HOME AND STARTING SIGNALS.

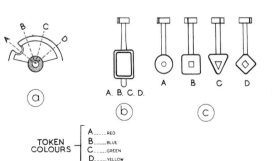

LEFT As has been mentioned, the driver must be in possession of a Token before he may enter a section. Tokens are one of four types – 'A' through to 'D' – and may also in some cases be combined with a different shaped end.

Chapter 3

Signals and Trackwork

Ever since the beginning of railways signals have controlled the movement of trains, telling them whether they were allowed to proceed or whether they should come to a stand. Trackwork was the rails on which the trains ran and points decided in which direction the train should travel.

Signals began in many forms, from Brunel's famous disc and crossbar to circular and semi-circular styles, but gradually evolved into the semaphore type, familiar in steam days. This was developed from Admiralty semaphore arms, placed on high points throughout the country, to warn of invasion by the French. By the 1890s the semaphore arm had become standard, with the arm in the horizontal position meaning danger, or caution, and in the lowered position, pointing downwards at 45 degrees, meaning clear. This type was known as lower quadrant. The majority of railways later changed to the upper quadrant style, where the arm was raised 45 degrees to indicate clear, because this avoided the need for heavy counterbalance weights to return the signal to

danger should the wire break, but there still is a mixture of both on the national rail network, and on some heritage railways. All signals are designed to 'fail safe'; in other words, if the wire is damaged, or breaks, the arm will return automatically to danger. They should also be capable of being lit at night and in poor weather. Originally this was done with oil lamps but some heritage railways have introduced low powered electric lighting, for the sake of convenience. In cases where the signal is in view of the signalman, he can observe the operation of the signal at night by looking at the aspect it displays, and where the back of the signal faces the signalman, the signal lamp displays a small white light to the rear, which is covered by a blind when the signal shows clear. Where a signal is not within the view of the signalman, he will have a repeater to show the position of the arm, either On, Off, or Wrong, if not cleared correctly. 'On' for a red signal means the arm is in a horizontal position indicating stop. 'Off' means the arm is in a 45 degree position indicating proceed, and 'Wrong' means that neither of these indications are displayed, and that the position of the arm is incorrect.

Trackwork also took many forms, from the Great Western's baulk road, where the rails were laid on longitudinal timbers, to others where the rails were laid on stone blocks. Problems maintaining the correct gauge caused there to be a rethink of these methods, and the wooden cross sleeper style became the most effective, with various styles of fastening the rail to the timber, usually a heavy iron casting, known as a chair, bolted to the sleeper, and a wooden key wedged between the chair and the rail to keep it in place.

Signals

Signals can be divided into two sorts – fixed signals, which are located at fixed locations, and hand signals, which are by their very nature movable, according to where they are required, or where they are given. Fixed signals can be further divided into running signals, which are sighted, observed and obeyed on the run; subsidiary signals, which are sighted, observed and obeyed, when the train has been brought at, or nearly to a stand, and movements are being made very slowly; and shunting signals, which are used for movements to and from sidings.

Running Signals

Running signals consist of two types. The first type is red with a white band on them, which are known as Stop Signals. Some people call them Home Signals, which is an incorrect term, because it is where the signal is sited that determines whether it is a Home or Starting Signal. Usually, Home Signals are on the approach side of a signal box, and Starting Signals are on the advance side of the signal box, but local conditions may mean that this is not always the case. Stop Signals have two positions – horizontal, meaning stop, and 45 degrees, either up or down, depending upon whether upper or lower quadrant, meaning proceed.

The other type of running signal is a Distant Signal, which is yellow in colour, with a fish-tail black band and a fish-tail end. It is the furthest away from the signal box, on the approach side, in order to give the driver the required braking distance for his train, and may even be motor operated. It gives two indications – the horizontal position, meaning caution, be prepared to stop at the next red signal, and a position either 45 degrees up or down, depending on the signal type, all clear, meaning that all the Stop Signals for the line to which they apply and which are worked from that signal box, are in the clear position. This means that the Distant Signal is the most important signal on the line, because it is the first indication the driver has of the state of the line ahead, and if he sees a Distant Signal in the clear position, he knows he can keep his train running at normal speed. If he sees a distant at caution, he knows he must bring his train

ABOVE There are not many preserved lines that can boast such a magnificent gantry (or, as they were once known, 'signal bridges') as here at Kidderminster, but it matters not whether it is one signal or a complex grouping, the principles are the same. Each of these signals, the four 'main' arms, four Calling On arms and three 'discs', are all STOP Signals and must not be passed until they are cleared. As seen, all may also be deemed to be in the 'ON' position. The train driver must know which signal applies to the line on which he is travelling – and also where he will go when the respective signal is cleared to the 'OFF' position. A basic point of note is that the taller the post, the more important the route. Both supporting columns also have white diamonds indicating to the driver that the lines are track-circuited and he therefore does not have to go to the signal box to carry out 'Rule 55' to remind the signalman of the presence of his train. The presence of the 'T' within the diamond refers to the fact there is also a telephone nearby allowing communication to the signal box visible in the distance. Should the signalman wish to contact the driver of a train waiting at the gantry, the external bells – on the side of the telephone cabinet – should be audible to the crew in the cab.

ABOVE A STOP Signal in the ON or danger position. Sometimes called HOME Signals, which is not strictly correct, as a STOP Signal can be the signal leading into the next section of line or the first signal at the end of a section. It may therefore be HOME or STARTING Signal, according to its location. This particular example is the down home signal for Bewdley South, obviously of the lower quadrant type, and seen in the ON and OFF positions. The white board is for sighting. Here the trees have recently been cut back, but against a background of foliage the position of the arm might not otherwise always be apparent, particularly in the autumn. The signal is wire worked, and the orange piping under the track contains electric cables to work equipment in the location cabinet alongside the post.

ABOVE The smaller silver box contains a telephone whereby the driver of a train detained at this signal may communicate with the signalman. (Track circuiting will also indicate to the signalman the presence of the train.) A sad fact is that the SVR S&T department made a beautiful little polished wooden cabinet, to contain the telephone. This was vandalised within hours of being fitted, hence recourse to the metal version seen here.

under control, and be prepared to bring it to a stand at the next Stop Signal. If when he gets to that Stop Signal, and it is displaying a clear indication, that only tells him that the line is clear, as far as the next Stop Signal, unless the signal concerned controls the entrance to the section ahead, in which case he can regain speed because the next signal he should expect to see, after the Section Signal, would be the Distant Signal of the box in advance.

Where there is a junction ahead, the Distant Signal applicable to the diverging route is always fixed at caution, to ensure that the driver reduces the speed of his train sufficiently to enable it to safely negotiate the pointwork, even though the Stop Signal at the junction for the route may show clear.

Having cleared his signals for the passage of a train, unless an emergency arises, the signalman should replace the Distant Signal when the train has passed it. But if a Stop Signal locks any points, that signal should not be replaced until the train has passed over the points that that signal locks, so that there is no possibility of the points being moved under the train.

ABOVE Signal and sighting board in a more appropriate setting. The telephoto shot does little to promote the state of the track – it really is not that bad.

LEFT A Distant Signal indicates the position of the next STOP signal. This particular example of a Distant Signal is 'fixed', meaning it will always show caution. Fixed Distant Signals are provided to warn the driver in advance of a place where he will always need to reduce speed – the entry to a loop for example. It will still have a lamp which must be lit during darkness. Notwithstanding the tree background, note there is no sighting board for this signal. This is because the driver is expected to know the line and will thus be aware of the placing of signals. The fixed distant might thus be said to be an 'aide-memoire'.

OPPOSITE LEFT A working distant in the ON, or caution position. This tells the driver that he must be prepared to bring his train to a stand at the next Stop Signal. The black box halfway up the post contains an electric motor to work the arm, and is used where a signal is a considerable distance from a signal box, and is operated by a shortened lever in the frame.

OPPOSITE RIGHT A working Distant Signal – when in the OFF position it indicates to the driver he will find all the STOP Signals relevant to that Distant Signal in the OFF position and so may continue at 'Line Speed'. (Line speed means up to the maximum permitted speed for the section of line concerned and/or his type of train.)

ABOVE A STOP Signal which is interesting on two counts. Firstly, it is one of the few surviving former GWR signals mounted on a concrete post. As mentioned earlier, the purpose of this book is to describe railway signalling operation rather than history, so suffice to say the GWR used concrete only for a short period although during that time several hundred signals were erected. More relevant is the positioning of the signal – on the opposite side of the line to what might be expected. Signals might be so positioned due to track or structural requirements and also when the view available to the driver would otherwise be restricted. In this case, the signal is one the side of a curve, and was so positioned to give drivers a better sighting of the arm, particularly those on GWR engines, which are driven from the right-hand side.

LEFT Bracket at Bewdley South. From this angle it may look as if a sighting board would be a good idea, but consider how this would then obstruct the view of the signalman in this direction. There are several points of note. Again the taller post indicates the most important route (into the down platform at Bewdley). The stop arm in the 'off' position and the distant arm in the 'off' position tell the driver that not only is he clear to enter the down main in the station, but that the next box ahead Bewdley North , has all his relevant signals in the clear position, and so the driver has permission to enter the section beyond the North box towards Arley. The Distant Signal referred to is of course controlled from the North box NOT the South box and is motor operated. To the right the lower post is for a diverging route, the Distant Signal here being fixed for speed and layout reasons. On the extreme left the small Stop Signal would indicate the route was set into the goods yard and not for passenger trains. Below the main platform is a small counter-balance Stop Signal. This is for movements into the 'Rock Siding' – better known as the locomotive yard. Note the lengths of the respective signal arms, 5ft, 4ft and 3ft. This is old GWR practice – the longer the arm the more important the route. Compare also the end of the distant arms where the shape is cut in what is referred to as a 'fish-tail', again old practice. The colour glass in the spectacle plate will be noted, this will be discussed later. Finally on the extreme left there is what is known as a 'Smoke Chute', intended to prevent damage to the actual bracket by the blast from the engine chimney.

ABOVE Now you know all about some of the basic signals . . ? So a test, why the different coloured finials at the top? Here the manufacturers wanted something not only to prevent water ingress into the post but sometimes to look decorative as well. On the GWR, red painted finals go with STOP Signals and YELLOW on Distant Signals. If a post has a mixture, as seen previously at Bewdley South, then the dominant colour will be RED. (Signal posts may be of wood, concrete or metal construction. In case of the latter, different railway companies used lattice or rails bolted together. Metal fittings would then be added.)

RIGHT & OPPOSITE Southern Region bracket signal at Harmans Cross on the Swanage Railway. These signals are clearly STOP Signals but this time of the upper-quadrant type – hence the different position of the sighting board. This is a lattice post, with preference to the right hand route. As the approach is on a single line it is clearly possible to have only one of these arms in the OFF position at any one time.

OPPOSITE Variations on a bracket – Swanage station home down signals. Clearly the top arm is to enter the platform, with a subsidiary arm underneath. The 'D' shaped box with the black disc in the centre is a route indicator telling the driver whether he is entering platform one or two. The ringed arm is for the run-round loop in the station and indicates that this is for non-passenger trains, and notice the white diamond on the substantial post as well as the ground disc, advising the driver that this section of line is track-circuited, and that the signalman will be aware of the presence of a train.

LEFT Hampton Loade up Home Signals. As there are no catch-points at Hampton Loade the locking is arranged so that only one train may enter the station at one time, the second train not being permitted to enter until the first has come to a stand. (As this is a regular passing station for trains, what this means in reality is that the driver will whistle upon his arrival at the Home Signal to announce his presence – the signalman then allowing whichever train has arrived first to enter the platform.)

LEFT In the previous view at Hampton Loade the signal was seen OFF for the loop. Here the signal box is switched out of circuit and long-section working is in place from Highley to Bridgnorth. All trains must therefore use the main line. The corresponding down Home Signal will also be OFF at the south end of the station.

SIGNALS AND TRACKWORK

THIS PAGE Bewdley North down inner Home Signals. These are what are referred to as centre balance arms due to restricted sighting under the station footbridge. That to the right is for the single line to Arley (Bridgnorth); the arm to the left now only refers to the short siding at the end of the loop but was formerly the signal for trains destined for the line to Tenbury Wells.

LEFT Bewdley North signals. As will be expected, the taller post applies to the principal route. In this case, the main platform is around the corner, not, as might appear from the photograph, straight ahead over the crossover. The lower post applies to a back platform behind the main platform in the station. Notice the difference in the distant arms, working for the main route, and fixed for the diverging route. Calling-on arms are provided for both routes, and when lowered, tell the driver that he is entering an occupied section, and must be prepared to stop short of any obstruction. The white diamond tells the driver that the portion of track he is on is track-circuited, the double disc Ground Signals are for shunting moves, and the '10' sign is a speed restriction through the station.

RIGHT Conventional lower quadrant Stop Signal. The worded sign advises the driver that from hereon there is a change in the method of signalling, in this case from double track to single track, and he requires a token to enter the section.

SIGNALS AND TRACKWORK

Subsidiary Signals

Subsidiary signals are a small arm, placed below a running signal, coloured white, with horizontal red stripes at the top and bottom of the arm. They give no message to the driver when horizontal and must not be raised or lowered until the train has been brought nearly to a stand. They may be one of three types, a 'Calling On' arm, which displays a 'C' when lowered, a 'Shunt Ahead' arm, which displays an 'S' when lowered, or a 'Warning' arm, which displays a 'W' when lowered.

Calling On means the driver is entering a section of line that is already occupied, and he must reduce the speed of his train accordingly, and be prepared to stop short of the vehicles, or train, ahead. Shunt Ahead allows the driver to proceed into the section, as far as is necessary, to clear the points, to enable him to reverse into any sidings for shunting purposes. Warning tells the driver that the line is clear only as far as the Home Signal of the box in advance, and he must be prepared to stop at that signal.

Some railway companies placed miniature arms beneath main arms for entry, or exit, to and from goods loops, and these were similar in design to red running signals, but much smaller. They were raised or lowered, when a train had been brought nearly to a stand and were required to enter a loop or non-passenger running line.

OPPOSITE Siding signal – hence the ringed arm – affording exit from a siding on to a running line. The Trap Point immediately beyond the signal will be seen to be 'open'. This is intended to deliberately derail a train/vehicle should it run past the signal at danger and so prevent it reaching the running line. The absence of a facing point lock indicates this is not a passenger line.

LEFT Lowered post Stop Signal with telephone attached. Notice the red and blue glass in the spectacle plate, not red and green as might be expected. Confused? Well think on – signal lamps were conventionally lit by paraffin which burns with a yellow flame. Yellow shining through blue produces green.

LEFT INSET Blue glass seen to advantage. The back of the signal lamp has a small hole, which seen from a distance displays a tell-tale light to the signalman that at night the signal lamp is burning and thus the signal is illuminated. When the signal is cleared on 'off' the shade at the rear moves and so obscures the light. The signalman can thus tell that the arm has responded to the lever.

RIGHT SR type lattice post containing a Stop Signal and Disc Signal. Neither signal is presently in use and to indicate this to the driver an 'X' has been affixed to both. As a temporary measure a Warning sign, 'STOP, obtain train staff or shunter's permission to proceed' has been affixed to the post.

OPPOSITE A subsidiary arm with a 'C' indicator behind it, meaning it is a Calling On arm. When the arm is horizontal, the letter is covered, but when lowered the 'C' is revealed to advise the driver that he is entering an occupied section. Naturally the indicator is lit at night by means of a lamp inside the housing.

SIGNALS AND TRACKWORK

RIGHT A backing signal, seen here at Bewdley, authorising movement in the wrong direction over the down line. (An alternative might be a ground disc.) Here the track in the locality will be seen to be track-circuited whilst the position of the 'cash-register' allows for more than one route to be selected: in this case either working in the up direction on the down main, or into the down siding. In the signal box two levers would thus be provided for this signal. Note the movement is only authorised as far as the next Stop Signal.

OPPOSITE If a stop (or distant) signal is situated in such a position that a driver would, when running at the maximum permitted line speed, have insufficient braking distance in which to bring his train to a stand once the signal comes into view, then a 'banner-repeater' is provided. (Should this repeat the indication of a Distant Signal then the arm seen would have a notch cut in the end.) As the name implies, this repeats the indication of the signal ahead. He does not have to stop at this banner-repeater.

Banner Repeating Signals

In cases where the sighting of signals is restricted, such as by bridges or curves, repeating signals were installed in order to give the driver advance notice of the indication of the signal ahead. They consisted of a black arm on a white background and were sometimes mounted beneath footbridges at stations or even on a post of their own. They gave no mandatory instruction to the driver, just repeated the position of the arm of the signal in advance so he could adjust the speed of his train if required, according to the indication given by the signal ahead.

SIGNALS AND TRACKWORK

Shunting Signals

These are generally placed near the ground and consequently are sometimes known as 'Ground Signals', or more colloquially as Dummies, Dollies or Discs. They are located at crossover roads, or points controlling exits from sidings on to running lines. They are worked from the signal box, and are usually circular in shape, painted white, with a horizontal red stripe across the middle of them. When turned through 45 degrees, they authorise the driver to

ABOVE Conventional ground disc or, as it is sometimes referred to, 'dummy' signal. The red bar indicates a definitive STOP. As seen earlier with the Siding Signal arm, notice the open Trap Point immediately beyond.

proceed as far as the line is clear.

Sometimes, the points from the sidings to the main line also lead to a headshunt, and in these cases the

ABOVE Ground Signals in the STOP position at Swanage. That to the right has a diamond which indicates track circuiting on the section of line to which that signal relates.

ABOVE Ground Signals in the OFF position. Note the casing above the signal in the lower illustration is a routing indicator, meaning this one ground signal affords access to more than one route.

disc may have a yellow band on it, not a red one. This allows the driver to pass the signal for the route for which it does not apply, such as if he wished to proceed up the headshunt for shunting purposes. If, however, he wished to use the route for which the signal applied, then the signal would have to be obeyed.

In cases where it is needed to use Disc Signals for two or more routes from a particular location, then the discs are mounted one above the other, and in those circumstances the highest one applies to the furthest left route, and the lowest one to the furthest right. Depending on the siting and design of the signal, four is generally the maximum number of discs one above the other, as they must give sufficient clearance for the trains and not foul the loading gauge.

Some Disc Signals are known as 'Running Dummies'. They are situated where there are points to a siding, or loop, in advance of a running signal, and it is necessary to pull the disc to a proceed indication before the signalman can clear the running signal. In these cases the disc allows the driver along the main line, and there is no need for him to reduce his speed, because the main arm controlled by the disc has authorised him as far as the next main arm, ahead of the disc.

LEFT In the previous view we saw the more modern approach to a signal affording access to more than one route. Left is the older, traditional, method where the two signals refer to two different routes. The simple rule where Stop Signals (which these are of course as they display a Red indication) is 'top to the left, bottom to the right'. Here at Bewdley South the signals apply to wrong direction running over the down line, in this case the line on the right. The top signal is for moves over the crossover, the bottom signal is for moves continuing along the right-hand line.

ABOVE The route is now set and cleared for a movement straight ahead on the right hand line.

ABOVE An older type of ground signal without a white backboard (the simple intention of the latter is to make the signal more visible). This is in effect a miniature semaphore.

ABOVE In appearance a conventional ground signal but this is in a fact what is referred as a 'Running Dummy' (Bewdley North No 10). It applies to all movements over the point, and unless this signal is cleared, it will not be possible to clear the main running arms applicable to this disc, hence the term 'Running Dummy'. In this particular location, the main arms are behind the photographer.

ABOVE An 'elevated disc' just to prove that a ground signal need not always be on the ground! The use of such signals was often for sighting purposes at the whim/preference of the local signal engineer/inspector or even more simply down to what equipment he had available!

ABOVE Less common, but still to be seen on a number of heritage lines, is the yellow ground signal, here seen at Swanage station. The driver may pass this signal when horizontal for the purposes of shunting on to a line for which it does not apply, such as a headshunt, but if the driver wished to use the route for which it did apply, in this example the crossover, then he would have to obey the signal.

RIGHT A banner repeater for a ground signal, at Swanage station on the Swanage Railway.

Signal Locations and Description

Taking the simplest of signal boxes, on double track, with a crossover between both running lines, there will be a red Stop Signal in advance of the box, known as a 'Starting Signal' and also as a 'Section Signal', because it controls the entrance to the section of line ahead. Then there will be a red Stop Signal on the approach side of the box, protecting the points on the crossover, which is known as the Home Signal. This will be followed by a Distant Signal the required number of yards away from the Home Signal, in order to give the driver sufficient braking distance to safely stop. The length of line between the outermost Home Signal and the Section Signal is known as 'Station Limits' and these limits are in fact movable, because the position of the relevant signals on one line may be different to those on the other line. There will also be Disc Signals between the running lines, for movements over the crossover, one for each direction.

As layouts expanded it became necessary to add further signals; for example, if points to sidings were required outside the Home Signal an additional Home Signal would be required to protect those points, outside the first one, so then the signals became outer and inner homes respectively. Similarly, with Starting Signals, there may in places be more than one of them and so they became starting and advance starting, but the signal controlling the entrance to the section would always be known as the Section Signal. All the signals and points will be interlocked, so that the crossover and Disc Signals cannot be reversed, with the running signals showing clear, and likewise, with the crossover reversed, the running signals cannot be put to clear.

In cases where the signal boxes are close to each other, the Distant Signals of one box may be on the same post as the Starting Signals of another box, and in these cases a mechanism known as 'Slotting' is used so that the distant cannot show clear, unless the Starting Signal also shows clear. If both signals are in the clear position when the Starting Signal is returned to danger, then the distant goes back to caution at the same time.

Hand Signals

Hand signals can have two forms – those given with a flag and those given with the hand, or at night with a lamp. Those given with a flag can have different meanings, depending on where they are given; for example, a flag signal given from a signal box can have a different meaning to a flag signal of the same colour given from a signal post.

If a hand signalman is appointed at a failed signal, or anywhere else out on the line, he should have with him detonators to place on the rail if required, and flags of the requisite colours to give the drivers the correct signal, according to the state of the line.

A red flag means danger, and should only be given

SIGNALS AND TRACKWORK

when it is necessary to stop a train, although in the absence of a red light at night, any light waved violently means danger. The exception to this is that a red light moved up and down vertically at shoulder height by the guard instructs the driver to create vacuum for the brake. A red light waved side to side at shoulder height tells the driver to destroy the vacuum.

A yellow flag can mean different things depending on where and how, it is given. For example, a yellow flag held steadily at the base of a Distant Signal tells a driver that the signal is defective and should be regarded as showing caution, but if it is given at a signal having a stop arm with a distant arm underneath it, it authorises the driver to pass the signal. On the other hand, a yellow flag waved side to side by a hand signalman tells the driver to reduce speed for emergency permanent way work, and having then passed that hand signal, when he then approaches a yellow hand signal held steadily, by another hand signalman, that indicates commencement of the speed restriction.

Green flags likewise can have different meanings at different places and times, again depending on how they are given, and when shunting a green light can have a different meaning to a green flag. There are many examples that could be given, because the rule book contains 14 variations of how a green hand signal should be given, but to quote just three: a green flag held steadily at the base of a signal carrying a stop arm authorises the driver to pass the signal; a green flag given by a signalman in his signal box authorises a driver to proceed, after having been brought to a stand at the box; a green flag waved side to side by a signalman, as a train approaches the box, tells the driver that his train is divided.

On the national rail network a green flag has largely been replaced by a yellow flag except in certain circumstances, and some heritage railways have also adopted this practice, so reference to the railway's rule book is required in order to understand which hand signals are used on that particular preserved railway.

Signals given by hand, not using flags, amount to six types: two arms raised above the head mean stop; an arm held horizontally, and waved up and down, means proceed slowly; one arm raised above the head means all right; one arm held horizontally, and moved in a circular manner, means move away from the hand signal; one arm held horizontally, and moved across the body, means move towards the hand signal; and one arm moved up and down vertically at shoulder height means create vacuum. Sometimes, an additional hand signal, a hand moved horizontally across the throat, is used, to tell the driver to destroy the vacuum.

Hand signals given with a lamp at night are mainly used when shunting, and also have several variations. A white light waved slowly up and down means move away from the hand signal; waved side to side means move towards the hand signal; and held steadily can mean different things at different times, according to who gives the signal. A white light given by someone on the platform, for example, tells a guard that the station work has been completed, but a white light held steadily by the fireman of a train acknowledges the green hand signal given by a guard. A white light twisted above the head by a shunter to a signalman tells a signalman that the train is clear of points that are required to be moved; a white light twisted above the head by a signalman to a shunter tells the shunter that the points have been moved.

A green hand signal given with a lamp held steadily above the head by a guard is well known as being the indication to a driver to start his train; but, when shunting, a green light waved slowly up and down means move away from the hand signal slowly; when waved side to side by a shunter it means move towards the hand signal slowly; but when given by a guard from the brake van of a freight train it tells the fireman that the train is complete, after setting off.

Whistle Codes

Some signals are given by drivers using the engine whistle. The list is quite long, and may vary on different railways, but one of the most common is three short whistles, meaning that a train or engine has been shunted clear of all running lines. Careful study of 'local instructions' is really the only way to get to know them all.

Trackwork

Plain track in steam days was made of a rail profile that was similar at the bottom to that at the top, a type known as bullhead. The idea was that the rail could be turned over, when the top face had worn, and be reused, but in practice that didn't work because of wear on the bottom face caused by the pounding of the rail against the chair. In addition, it was expensive and time consuming to maintain all the fittings needed for that style of rail so a more modern type was developed, called flat-bottom, which had a much wider web at the base, and was clamped via a rubber pad, direct to the sleeper. Rails were made of different weights, increasing as train weights increased, until bullhead was standardised at 95lb per yard. Flat-bottom was heavier, at 109lb per yard, increasing to 112lb today. With fewer fittings and heavier weight it lasted longer, and required less maintenance.

Bullhead rail was laid on wooden sleepers, but flat-bottom is nowadays clipped to concrete sleepers because a wooden sleeper will last a lot less than a concrete sleeper which has a life expectancy of 25 years. The common rail length for jointed track was 60ft but as welding techniques improved they became 300ft lengths, and on the national rail system, have been welded in to even longer lengths. Track lengths of 60ft are joined together by fishplates, which are metal bars about 18in long fitted on each side of the rail, and clamped together through the

ABOVE Here is seen an 'FPL' (Facing Point Lock) but minus a protective cover. The tie bar connects the blades of the point ensuring they move simultaneously. When the signalman pulls the lever for the FPL (coloured blue) the 'bolt' will move forwards into the slot in the tie-bar. In this position the point may therefore be said to be 'locked'. The point is then safe for the passage of a passenger train in the facing direction. The relevant point lever in the signal box (coloured black) cannot itself be moved until the FPL is released. (Note – if a passenger train is required to pass in a facing direction over a point which either is not fitted with an FPL or has a defective FPL, then the point MUST have a clip and padlock fixed to it.)

rail, usually with four bolts, although some railways in later days developed a two-bolt fishplate for economy of fixing.

Where a portion of track had a small electric current running through it, known as a track circuit, the fishplate had special plastic fittings placed between the metal bar and the rail, and also between each rail end, known as an insulated rail joint, in order to confine the length of track through which the electric current was required to run. Rail joints within a track circuit that were not required to be insulated were bonded by two lengths of wire, bridging the fishplate and maintaining electrical continuity. Track-circuits are normally placed in rear of outer home, or advanced starting, signals, or where sighting is restricted, to indicate to the signalman the presence of a train. Where they are located, a white diamond plate is attached to the signal post to tell the train crew that there is no need for one of them to go to the signal box to tell the signalman that there is a train stood at the signal.

Track laid on curves was given a super-elevation, or cant, by packing the sleepers on one side of the line with additional ballast. The degree of cant was calculated according to the sharpness of the curve and the speed of the trains. It was very precise and transitional, increasing as the train entered the curve and reducing as it left the curve because that gave the train a smoother ride.

Pointwork

Points come in many types and formations, and are worked by heavy metal rodding running out from the base of the signalbox to the required location. In the days of the private railway companies this rodding was round in profile, but in British Railways days a substantial square 'U' section was used, although both types are still in existence. The rodding runs, as they are known, can be quite long, and from initial cranks just outside the signalbox will run parallel to the track until they reach the position in which they are needed to work, where they are linked up to the point switch-blades. At about half-way along the run compensator cranks are positioned which convert a pull to a push, and take up any expansion which may occur in warm weather. Points are said to be facing when they change the direction of travel of the train, and trailing when used in the opposite direction. When the diverging route goes to the left, they are said to be left-hand, and right-hand when it goes to the right. The diverging routes can be of many angles, according to the speed necessary for the trains, and are generally referred to as 1 in 8, or 1 in 5, for example.

Terminology for points is quite complicated, but in simple language the rails that alter the route of the train are known as the switch-blades. They are positioned within the running rails and that area of the point is called the toe. The switch-blades lead on to the closure rails, which in turn lead on to the crossing, which is

LEFT A trailing point at Bewdley, showing the exit from the goods yard on the left. Notice the running signals in the distance, indicating that the normal direction of travel is straight ahead, but that there is also a backing signal, with two holes in the arm, for movements, if authorised, in the wrong direction. The right-hand picture is a closer look at the same point and shows that for trailing points, no point lock is necessary, but if a passenger train was to be run over the line in the wrong direction, a clip would have to be applied to the point blades.

V-shaped, sometimes known as a frog, although no-one knows for certain why. The crossing has wing-rails either side of it, and on the opposite rail are check rails, all to help guide the wheels smoothly through the junction to where the two routes have finally separated, which is known as the heel end of the point.

The switch-blades are joined together by stretcher bars, probably three in number, spaced along the blades, but the one at the toe end of the point will have a notch cut in it to accommodate a lock, which is a flat bar plunger which goes into the notch to prevent the blades moving under the train. Some blades are locked in both directions – normal for straight ahead or reverse for the diverging route – and some only in one position, depending over which route it is required to run passenger trains, but it is a government inspectorate requirement that all routes over which passenger trains are run are equipped with locks and the permitted gap between the running rail and the switch blades, when closed, is checked with a special feeler gauge to ensure it is within limits. Earlier locks were linked to a bar, which ran along the inside of the running rail and was known as a 'Fouling Bar', to prevent the signalman altering the points while a train was stood on them. In later years, this function was taken over by track-circuits, which electrically lock the relevant lever in the frame in the signal box to prevent it being operated, and so unlocking the points.

Detection

All points linked to signals must have some form of detection, to prove that they have been placed in the correct position, before clearing the signal, because clearing a signal proves the route to which that signal applies. In mechanical form this is a round bar, bolted to the switch-blade. This round bar is in turn bolted to a flat bar, placed on its side, which has a notch cut in it. When the point is moved to the correct position the notch lines up with a slide, at 90 degrees to the bar, which is connected to the signal wire. When the signal lever is pulled, the wire pulls the slide through the notch and the signal is cleared, so proving the route. On later installations, the detection is proved electrically in a box alongside the point blades and the signal lever is locked in the frame until all the correct contacts in the box have been made.

ABOVE Basic detection. 'Interlocking' is what applies to levers in the signal box Lever Frame, meaning a lever cannot be pulled if it would set up a conflicting movement. 'Detection' is on the ground and is used to prove that a route has been correctly set, before the signal can be cleared. In these two illustrations, notice the round tie-bar attached to the end of the point blades, and a flat bar with a notch in the bottom at the far end, attached vertically to the right-hand end of that arm and running through a casting on the ground. In the right-hand illustration, the point blades have been pulled over, taking the round bar with them and consequently the flat bar as well, so lining up the notch with the wire and slide to work the signal.

ABOVE Detection in detail. As before, the horizontal slide (from the point) must be in the correct position before the vertical slides (attached to which is the signal wire) may be pulled through it. The notch in the slide can be seen at the bottom of the bar, to the right of the casting. When the point is pulled over, this notch should line up with the vertical slide attached to the signal wire, enabling the signal to be cleared. If for any reason the alignment is not correct, then the signal will not clear, because the route will not have been proven.

ABOVE For a passenger train to pass over a facing point – 'facing' means it has the option of being diverted on to two or more routes – the point MUST be locked. This is achieved by a similar means of detection (basically a bolt that aligns into a slot). To ensure smooth and accurate movement of the mechanism the whole is kept well oiled and as a result will sometimes be covered by a plate when not needed to be exposed for maintenance. Note the fouling bar inside the running rail.

ABOVE Here we see a TRAP Point and the rear view of its associated ground signal. A TRAP Point is deliberately intended to derail a vehicle or train should it run away before it reaches a running line. Often there will be just a single blade and, it will be noted, no FPL. (Note the difference between a TRAP Point and a CATCH Point. A TRAP Point is as previously described, a CATCH Point will be located at intervals on and at the foot of an incline, intended to deliberately derail a train or vehicle which runs back the wrong way down an incline. These are invariably spring-loaded and have no effect on trains passing in the right direction.)

Trap and Catch Points

There is always some confusion over the difference between Trap and Catch Points. Trap Points are used at the exits from sidings and consist of a break in the running rail and a single point blade, worked from the signal box, generally accompanied by a Disc Signal. They are placed there to prevent any runaway vehicles from going on to the main running lines. Catch Points are sprung, and are not worked from a signal box, and

are placed at the bottom of steep gradients to prevent any breakaway vehicles from heading backwards in the wrong direction of a running line. They were very common on the national network in the days of loose-coupled freight trains, but now that all freights are fully-fitted they have largely disappeared from the scene. They are not used on single lines because of the bi-directional nature of the trains, and the impracticality of closing them up.

ABOVE AND LEFT An unusual, but necessary, double facing Trap Point. The location is at Cranmore on the East Somerset Railway where the private line meets the National Network. A movement from either direction requires the consent of the other.

Advanced Pointwork

For more complicated layouts, or where space is limited, advanced trackwork is used, and can take several forms. Sometimes there will be a diverging route to the left, and the right, combined in one point, known as a three-way tandem; sometimes both diverging routes will be to the left, or the right. Where one track crosses another on an angle, and there is no way of running from the angled route to the straight route, this is known as a diamond. Where it is possible to go from the angled route to the straight route, on one side, this is known as a single slip, and where it can be done on both sides, it is known as a double slip. These track formations are only used where absolutely necessary, because the associated signalling works are very complicated and need constant and careful adjustment.

ABOVE Points and locks that are mechanically operated by a lever away from the actual location of the point/lock will have this movement transferred by rodding, in itself either round or 'U'-shaped in section. To allow for expansion and contraction, compensators are provided at intervals. Note above the compensator is a run of signal wire.

OPPOSITE & LEFT An outside ground frame. A ground frame is a lever(s) used to control points/locks but which is not a controlling signal box, meaning a GF will just control equipment in a very local facility. There are numerous types, the examples seen each being of two levers, one blue and one black, meaning an FPL and a point. (As the name implies, a ground frame is usually on the ground, but it does not have to be in the open, and the levers might well be in a hut where there might also be a telephone to communicate with the shunter or signalman. A former signal box could also be downgraded if no longer a block-post and so take on a new use as a ground frame.)

OPPOSITE PAGE A single point lever but with an Annett's lock fitted. This lock holds the lever in a set position until the relevant Key is inserted. The Annetts Key may not be released until the lever is restored to its normal position. (The white paint on the top of the handle is purely to assist in identification at night and also to prevent rusting.) On the lever the white plate (lever 'lead') describes what is required to unlock the lever before it can be operated. The lock is, as might be expected, painted blue and is where the Key to release the lever will be inserted.

LEFT A slightly more complex outdoor ground frame (at Midsomer Norton) including: one FPL, two points and three Ground Signals. The GF is released by inserting a Key on the train Staff.

ABOVE Hand point lever. This one is located in the 'six-foot' (between the two running rails) and in consequence folds down when not in use.

INSET A rail 'chock' – again different designs are in use. This would take the place of a Trap Point. Sometimes a rail chock might be remotely operated by lever, other times it was a permanent or temporary fixture. When in use it will be padlocked to prevent unauthorised removal.

TOP Track circuiting. Two wires joining connecting sections of track. The fishplate will also afford some continuity but is not relied upon, hence the bonding. (Should the fishplate mark the end of a track circuit then insulated sleeves will be added to ensure a break in any electrical supply.)

ABOVE Some sidings, not running lines, may have stone blocks and metal ties at intervals. This method is historic and applied to a time when timber may have been in short supply.

ABOVE Gradient post. The '88' refers to a drop of 1 in 88 (feet). Any rail vehicle(s) left on a gradient MUST have the brakes fully applied.

LEFT The commencement of a 10mph speed restriction. The driver must ensure the complete train has passed over the stretch of line subject to the speed limit before he may increase speed.

Tunnels, Gradients, and Section Times

All signalmen will need to have knowledge of the geography of the railway on either side of them and how long it will take a train to traverse that portion of line. A tunnel, for example, means certain precautions have to be taken, should it be necessary to carry out an examination of the line using a train. Passenger trains are not usually permitted to be used for such a purpose, in case the obstruction is within the tunnel. An engine on its own is the best means of carrying out such a procedure, but if that is not available the line may have to be examined by someone on foot.

Gradients are important, not only for calculating section times but also in case a freight train should

LEFT A mile post – again different companies used different designs. The mileage will usually be from London – although on some branches and cross-country lines it could be from the actual junction. Here the distance is 137½ – the two vertical bars indicating '2 x ¼' mile.

located. Mile posts are marked in quarter-mile intervals, and the number of a mile post is particularly important when advising train crew or any members of staff where they need to look for anything unusual.

Section times need to be known, especially on single track, for the regulation of the passage of trains. This is very useful when trains get out of course on single lines, where trains cross, because then the train crew can be informed of the expected arrival time of a late running train and in turn, the passengers can be advised of the likely length of any delay. It is also worth being aware of the running times between other signal boxes along the line, so that calculations can be made for shunting moves or deciding which trains should have priority over others. Another useful idea is to have in mind the driver's view of signals, as this can be particularly useful when making propelling moves during shunting, especially where the line is curved and the view is restricted by the rolling stock ahead of the driver. It can also be a help in regulating trains to know how long after leaving a signal box in rear it would be, before a driver is within sight of the outermost signal he is approaching.

break into two or more portions. In this event the gradient would determine in which direction the runaway vehicles would be likely to travel. Action should be taken to deal with any runaways if they arrive back at the signal box or if there is another train following. A slow moving freight train will take longer to travel through a section up a steep gradient than a passenger train will, and likewise a loose-coupled freight train will also have to proceed slowly down a gradient, remembering that the train will only have brakes on the locomotive, and rear brake van, in order to prevent the whole train running away. All signal boxes should have a chart showing the gradients of the line, together with any engineering features like tunnels or viaducts and all the relevant mile posts, their number and where they are

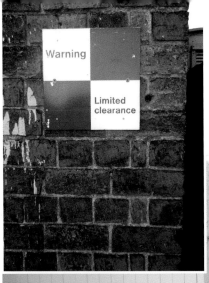

LEFT & BELOW Limited clearance markers are provided where there is insufficient space for a person to stand should a train be passing or about to move nearby. Here the location is alongside Bewdley Goods Shed with the danger all too apparent.

Chapter 4

The Signal Box

When railways were first built in the 1840s policemen were stationed at the signals and rotated the signal post according to the state of the line, which told the train driver whether or not he could proceed and at what speed. Even today, the nickname for a signalman is a bobby, after the founder of the Police Force, Sir Robert Peel. As the railway network expanded, and more men were required, this became too heavy a drain on police manpower so policemen were replaced by pointsmen, whose duty it was to set the points and signals for the direction in which the train was to travel. This entailed walking to one set of points and setting them in the correct position, then walking to the next set, doing the same again, and so on, until all were correctly set, and then the signals were placed in the correct position to give a proceed indication to the driver of the train.

All of this meant a lot of exercise and there is a story that one pointsman acquired some long lengths of rope and tied them to the points under his control, and just pulled on the ropes to alter the route and the signals. One of his colleagues reported him to his superiors, as irregular working, but the railway company thought there might be something in this idea as all the equipment could be worked from a single location and fewer men would need to be employed, and so the signal box was born.

The Signal Box

Signal boxes come in all shapes and sizes; some are of timber construction, for economy and ease of erection. Some of the more substantial ones are made of brick, and some are a mixture of both, and no two boxes are identical because each location is geographically different and each will have its own set of special circumstances. All of them, however, contain the equipment necessary to signal trains, and that equipment can be divided in to four basic items. The first is an instrument of some kind, depending on the type being used, containing either a Staff, a Token, a Tablet or a Key, which has to be released to give to the driver, as his authority to enter the single line, and proceed to the next signal box. The other items are a Diagram, a Block Shelf and a Lever Frame.

ABOVE, OPPOSITE PAGE AND OVERLEAF To become a signalman you will of course need a signal box. The temptation can be to say the larger the box the more complicated the working and that may very well be true in some cases. However, in others, one lever may be king to several types of moves.

THIS PAGE , OPPOSITE PAGE AND OVERLEAF Signal boxes come in all shapes and sizes, according to the era in which they were built, and the size of the frame inside. Some were very substantial, and made completely of brick, such as this one at Kidderminster on the Severn Valley Railway (opposite), and the one at Cranmore on the East Somerset Railway (above). Others were half brick and half timber such as the one at Harman's Cross (left), while yet another variation was timber on a brick base wall on a platform seen at Corfe Castle (page 87), both on the Swanage Railway. The simplest method of construction was all timber and free standing such as the one at Minehead on the West Somerset Railway (page 86).

THE SIGNAL BOX

The Staff, Token, Tablet or Key

If the railway is single track the driver has to carry with him some form of authority to be on the line and to prevent other trains being on the line at the same time. This authority can take many forms, depending on the type of signalling being used on that particular railway. Some railways, like the Great Western, had their own signalling department and made everything themselves, whereas other railways purchased their equipment from outside signalling contractors. The simplest form of instrument contains a wooden Staff with the name of the section of line to which it applies carved on it, and a metal Key attached to the end. The Staff is then turned through 90 degrees in the instrument, and this provides an electrical release to the signal lever, controlling the entrance to that particular section of line. Once the lever controlling the signal has been pulled, and the Staff removed from the instrument, the lever cannot be pulled again until the Staff is returned to the instrument, in order to prevent another train being on the single line at the same time as the first one.

ABOVE So, to train operation and control on a single line. The most common method used is that of a Token which is released with the consent of two signalmen – one from where the train will depart from and the other at the end of the section over which the train will pass. Seen here are two different types of instrument, nearest the camera is a TABLET instrument, whilst the red coloured machine is a TOKEN instrument.

ABOVE LEFT Two different uses of a Token machine. Both are of the older GWR style, the 'yellow' Token being short-section working from Bewdley North to Arley, while 'green' is long-section from Bewdley North to Highley. The knob on the top of the machine is colour-coded to the Token it takes.

ABOVE RIGHT A legitimate alternative to a Token to authorise a driver to be on a single line is 'Acceptance Lever Working', where instead of Tokens, there is a special lever in each signal box at each end of the section, and between the two signalmen, only one lever can be pulled at a time. When the signalman at the far end of the section pulls his lever, an indicator at the other end shows the signalman there that the lever has been pulled, and a train accepted. Seen here, an up train has been accepted by the Kidderminster signalman.

LEFT Double line working, in the form of a Spagnoletti Instrument, so called after the name of the designer. The upper half shows a red indication in a white panel, meaning there is a Train on Line, while the lower green panel shows a split white/red indication meaning there is no train anywhere in the area on the down line.

ABOVE The Signal Box Diagram indicated the layout of the lines under the control of the specific signal box. It is drawn so that points are shown in the position they stand in when the controlling lever is 'normal' – in other words, not pulled over – in the frame.

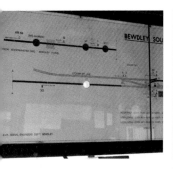

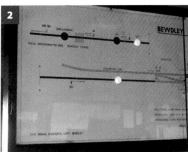

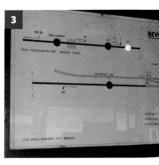

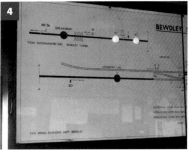

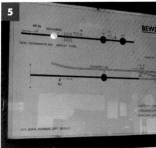

The Diagram

The diagram is a plan of the area controlled by the signal box. It is usually hung from the ceiling, and encased in a wooden frame. It will have marked on it where all the points and signals are, and numbers showing which lever works which point or signal, the levers in the frame generally corresponding with the position of the equipment on the ground. Some diagrams will have solid black lines representing the track, with small bulbs in them, which become illuminated when a train stands on that particular portion of the railway line and so indicates to the signalman the presence of a train at that location. Sometimes the bulbs are white, sometimes they are red, depending upon what signalling era is being

ABOVE Here we see the progression of a train on an illuminated track diagram (Bewdley South). The service has departed from Bewdley and is en-route to Kidderminster. Track circuits are of variable length and dependent on the length of the train; more than one may be illuminated at one time.

represented in the box. This is known as a track-circuit indication, and is achieved by feeding a low voltage current along the rails, usually no more than 6 volts.

THE SIGNAL BOX

When a train runs over the rails, it completes the electrical circuit and the bulb lights up. Lines that are not track-circuited are drawn on the diagram either in pale blue or grey.

The Block Shelf

The Block Shelf is a substantial piece of wood, again hung from the ceiling, below the track diagram and running the length of the Lever Frame. It has mounted on it various instruments and indicators to show the signalman the state of the line, position of the points and signals, whether or not the signal lamps are lit, and if working to other signal boxes, a bell with a morse tapper underneath it for sending and receiving messages from signalmen in signal boxes either side of him. If the diagram does not have track-circuit indications, the Block Shelf may have special indicators to show the presence of a train, should any part of the line within the signalman's control be track-circuited.

ABOVE At some locations where only limited Track Circuiting is in use, instruments are used instead – again in varying designs and in this case colours. When the track is clear the indicator rests at an angle of 45°; when occupied the bar takes on a horizontal position. (As the signalman becomes more experienced he/she can tell by the very slight click that accompanies the change of position of the indicator which track is occupied.)

THIS PAGE To communicate with the signalman on either side, bell communication is used. The code for a stopping passenger train for example is '3-1', which is tapped out as '3 pause 1'. Signalmen get used to the speed of tap from the man on either side, although if too fast the solenoid is not given time to operate and all that is received is a series of clicks. Bells are deliberately made to a different pitch – there is also a gong – so that the signalman immediately knows which signal box is calling. (On a single line the bell tapper is a push button plunger on the single line Instrument – Tablet/Token machine.) For double line working the tapper is usually attached to or in the proximity of the associated bell.

ABOVE LEFT & CENTRE If a signal cannot be seen from the signal box then it is the practice for an electrical repeater to be provided to confirm the position of the arm. In the left view the indicator shows the signal at ON (danger) – the lever being normal in the frame. In the right hand view the lever has been pulled and the arm has responded to the OFF position. Note, if the indicator moved to the WRONG position it would mean the arm has not made a fully ON or OFF position. The signalman must observe his indicators and in such a case 'restroke' the lever – meaning restore and re-pull the lever. Mechanical signals can be up to one mile from a signal box and a lot of effort is required. The number '30' refers to the number of the signal in the frame.

ABOVE RIGHT Circular brass cased instrument repeating a Distant Signal. (Again, different companies used different designs of equipment.)

RIGHT An early type of repeater for a Distant Signal. This one shows its age as the miniature arm is painted red – this colour ceased to be used for Distant Signals on the GWR from 1927.

LEFT TOP, LEFT BOTTOM & RIGHT Points are often indicated in the same way. 'N' or NORMAL being their position when the lever is normal in the frame, 'R' or REVERSE when the lever has been reversed. Again the point(s) concerned will be identified. WRONG means the blades have not responded correctly.

BOTTOM RIGHT Emergency release for a set of points that are otherwise locked by a track circuit. To activate the button behind the piece of paper the glass seal needs to be broken. This is a last-resort operation and NEVER something to be carried out in the normal course of events. (It would clearly warrant explanation in the Train Register.) If, for whatever reason, instruments or levers are not responding as they should the best recourse is STOP and THINK. Better a train is delayed than an accident or worse occurs.

THE SIGNAL BOX

BELOW Signal and other indicators; also on the Block Shelf are a number of release plungers, again with their respective lever number. If related to a signal, the plunger must be depressed momentarily, at which time the related electric lock should be heard to 'click' (pick-up); the plunger may then be released and the signal lever pulled. It is not necessary to press the plunger to restore the signal. If related to a point, then the plunger needs to be kept depressed until the lever is out of its position in the frame in either direction – on this occasion the plunger is the check to ensure that the relevant track circuit associated with the point is free from any vehicle.

ABOVE A 'Lamp Indicator' to show when the signal lamps are lit. This may indicate the lamps in several signals. If a lamp fails a bell will sound to alert the signalman. In this illustration, the red indicator shows a lamp out. If all the lamps are lit, the indicator will switch to green and show Lamp In.

If the signal box can be switched out of circuit, there will also be what is called a Block Switch somewhere on the shelf, usually in the middle and painted red with a brass handle on it, which is turned through 90 degrees to either make or break the electrical contacts between the adjacent signal boxes, depending on whether the local box is required to be in circuit or not.

The Lever Frame

The Lever Frame houses all the necessary levers to work the points, signals, facing point locks, and, if appropriate, to release or lock any crossing gates. Usually, the frame faces the track, and the levers are numbered from left to right, but there are exceptions, depending upon the geographical location of the box, and these are known as 'back to traffic' boxes.

The colour the lever is painted indicates its function in the frame. Red levers work Stop Signals, Yellow levers work Distant Signals, black levers work points and blue levers are for facing point locks. Some red levers may have a 2in-wide white band painted on them about half-way up the lever. This is to indicate that the lever is electrically locked by whatever type of instrument is used to control the single line, as that lever works what is

LEFT The first thing to be noticed on entering the signal box (apart from what should be care taken by the duty signalman over cleanliness – NO dirty boots *please*) will be the actual Lever Frame. The traditional method had been for the frame to face the traffic, but some boxes in recent years have had the frame to the rear of the box. The thought behind this being that the signalman will then have an uninterrupted view of the train as it passes.

THE SIGNAL BOX

RIGHT Levers are painted different colours to identify the type of equipment they operate. RED for a Stop Signal, YELLOW for Distant Signals, BLACK for points, and BLUE for locks. Spare levers are WHITE. Sometimes a particular lever will display more than one colour – as here, 'red with a white stripe'. In this case it means lever No 5 'Down main advanced starting' signal can only be pulled when 'Line Clear' has been given from the box in advance.

LEFT A 'lead' will be provided on the lever to remind the signalman of the operation of the particular lever and to identify which levers must be pulled first. In the case of No 4, lever number 7 must be first. And in the case of No 8, levers 7, 19 and 17 must be operated in the order shown before No 8 may be pulled. Note when replacing a lever the reverse order applies. The locking will always allow a signal to be restored to ON at any time, but under normal working this must only ever be done when the train has passed clear of all connections locked by that lever.

called the Section Signal, which is the last signal the driver will see before he proceeds towards the next signal box. Other levers may be painted half red and half black, and these levers work signals on relief or loop lines. Any levers that are painted white are not used, and classed as spare in case they are needed in the future for additional points or signals. If there are crossing gates worked from a signal box, these are usually locked or released by brown levers. All levers are interlocked so that the route cannot be set for one direction and the signals cleared for another direction, and likewise neither can conflicting routes be set up or signalled. This means that the levers have to be pulled in a particular order, and each one has what is called a 'lead' fixed to it, which is a metal or plastic plate bolted to the front of the lever, showing which levers are required to be pulled before the one in question will be free. Some frames have only one number on each lever, which then tells you the one to pull after that. For example, lever 6 may have an 8 on the front, so that lever is required to get number 6, but lever 8 may have a 7 on it, so that is required to get number 8, and so on. There is a knack to pulling each individual lever. Some can be very heavy as points can be 300yd or more away from the signal box, and a signal which might be right outside the box may have a long wire run and so also be very heavy to pull. Levers are generally arranged in the frame so that those operating the signals are at each end, signals on the left working signals sited for trains approaching from the

ABOVE The respective levers will operate the equipment outside the signal box; in this case two separate levers will operate the two separate Stop Signals. (The distant is fixed.) The finial corresponds to the colour of the signal arm.

Lever leads at Swanage.

16
GOODS
SHED
POINTS

17
FROM
GOODS
SHED
SHUNT
13, 16

18
ALONG
No.1
ROAD
SHUNT
13

19
ALONG
DOWN
MAIN
SHUNT
EAST

20
FROM
ENGINE
SHED
SHUNT
21

21
ENGINE
SHED
POINTS

ABOVE Lever leads at Swanage.

LEFT The interior of Hampton Loade box. When a lever is in its usual position in the frame, the lever is said to be NORMAL. When it has been pulled across, it is REVERSED. Here Nos 1 to 4 are normal and No 5, the FPL, has been reversed. Behind the top of the actual lever handle is the 'catch'; this is pulled and releases a notch in what is known as the quadrant-plate. The catch is pulled along with the lever and the former released as the lever reaches the full extent of its travel; the catch will then drop into another notch at the opposite end of the quadrant plate to hold the lever in the reverse position. Restoration is the opposite of that described.

left of the box, and levers on the right working signals sited for trains approaching from the right. In the centre will be the levers working the points and point locks, probably with a gap of two or three spaces between those levers working the signals, to allow for any additions to the frame which may become necessary in the future. Some levers have a short handle; this indicates that the point or signal worked by that lever is electrically operated and so does not need to be pulled very hard.

The Train Register

Apart from the Single Line Instrument, diagram, Block Shelf and Frame, there are various other ancillary pieces of equipment required in a signal box, the most important of which is the Train Register, in which all the times of bell codes being sent and received, all passing times of trains and any other unusual occurrences are recorded.

BELOW Ask what is the most important item in the signal box and the surprising answer will be the 'Train Register'. Here every movement is logged, times and descriptions as well as the detail of any out of course occurrence. In the event of any incident it is the Train Register that will be minutely scrutinised.

THE SIGNAL BOX

ABOVE Associated with the Train Register is the signal box clock. Traditionally a 'time' signal was sent every day and checked against the signal box clock. (On the GWR/WR this was via the omnibus telephone at 11.00.) Should the clock be seen to be fast or slow, the amount is noted in the Train Register and the clock also adjusted accordingly.

RIGHT TOP & BOTTOM Most signal boxes on preserved lines will have a power supply for the signal box and its equipment – in days of yore this might well have been batteries – but we have moved on since then. Accordingly the first task for the duty signalman will be to turn this supply on – different arrangements will apply locally.

ABOVE Some signal boxes may be switched out of circuit at times, the 'section' being referred to as 'Long' or 'Short Section Working'. Shown here is a 'Switching Out' instrument – a simple turn of the brass lever to position 'in' or 'out' – meaning box in circuit or out of circuit. Inside the instrument are a number of contacts which when turned to 'out' will re-route the bells and certain instruments, so by-passing the particular signal box. Note in this example a locking pin is provided to prevent accidental operation. (Other designs of Switching Out Instrument are available!)

It consists of a double-page spread of columns, with a row of headings at the top, listing where the times of the bell codes should be entered for both the section in the rear and the section ahead, and in which direction the trains are travelling. Down train times are entered on the left-hand page, and Up trains are entered on the right-hand page. The Down direction is always away from London, or the main-line station, and Up trains are always towards London, or the main-line. The register is a legal document in the event of there being any inquiry, and is regularly checked for accuracy by either the Station Master or signalling inspectors.

LEFT An oil or electric handlamp is essential. Note: if you are working other than in daylight and reliant upon an oil (paraffin) lamp, this MUST be immediately ready for use – meaning it must be lit! In either type the glass must also be turned to display the RED indication. In an emergency seconds count.

ABOVE RIGHT Another essential item is a set of flags. RED clearly means STOP; the use of a GREEN flag held steady from the signal box can have a different meaning than might first be intended. We will not go into great detail here, although suffice to say in the days before Distant Signals and their associated lever were yellow, a little rhyme was used as an aide-memoir, 'White is right and red is wrong, green means gently go along'.

Ancillary Equipment

Other items in the signal box are telephones, for communication not only with other signal boxes but other departments and stations on the railway; detonators, for placing on the track for use in an emergency, and to protect anyone working on the line, or to protect a failed train; wire adjusters, to absorb or remove any slack in signal wires, in order to get them to show the correct indication; lever collars, which are placed on a lever to remind a signalman not to pull it; a gradient profile, which illustrates where the railway line rises and falls, and by how much; and a clock for noting times. There should also be a set of flags, red and green, possibly yellow also, and a desk on which the train register is placed. Inside the desk are

kept the Rule Book, the book of Train Signalling Regulations and a Sectional Appendix. The Rule Book tells a signalman what he must do, the Regulations tell him how he must do it and the Sectional Appendix gives details of Local Conditions and restrictions along the railway. These will be discussed in a later chapter. Most signal boxes have special conditions, which apply just to that box. These are known as Footnotes and are generally posted on a large board fixed in a suitable place to the wall of the box. A working handlamp, either of the oil-burning or battery type, and capable of displaying red, yellow and green aspects, should be in a prominent place in the box, generally on the Block Shelf. This is for use in an emergency, should it be required to stop a train.

THE SIGNAL BOX

LEFT Signalmen always do it with dusters; it's a 'badge of office' and you will be pulled up very sharply if you don't use one. The reason is to keep any perspiration from the hands off the steel of the lever to prevent them rusting, which would in turn give the signalman blisters.

BELOW What is it, and why (when you have worked it out) is there an illustration of a window catch? Simply remember to release the catch on the window so it may be opened as soon you start duty – like having the handlamp ready for use. This is to ensure a message may be heard, or passed without delay. (Signal boxes, with their large glass area, can also become very warm in the sun!)

LEFT Personal comforts!

THE SIGNAL BOX

Telephones

The signal box is likely to have several telephones. They may be of the dial type, directly connected to a signal for the train crew to contact the signalman, as well as what is known as an 'Omnibus' phone, whereby all the signalmen can talk to each other at the same time. The omnibus phone has buttons on it which are pressed to call another signal box. Each signalbox has its own special code, for example one long and three-short, so that the intended signalman knows if he is required to answer the phone. Some communication is now being done by radio and in these circumstances the NATO phonetic alphabet is used where necessary.

ABOVE Telephone communication between the various signalmen is used to clarify an issue or explain a peculiarity. For example, if a train were seen to pass with a door handle not properly turned, the signalman would immediately send seven-beats on the bell to the signal box ahead and then speak to the man on the phone to explain the reason for the code – which door on which carriage etc. Under these circumstances the next course of action will depend on the reply received back: if all is in order then other trains may follow on to the same section of line, if however there is a chance that someone or something has fallen from the train, then no trains may use the same section of line until it has been examined. Telephones can be restricted to box to box only, box to station, 'omnibus' (meaning anyone may listen in), and of course conventional BT type network.

Detonators

Detonators are small explosive devices, about 2in round, with lead strips attached to them so they can be clipped to the rail. They give a very loud bang when a train runs over them, in order to attract the attention of the train crew. They have a life expiry date, usually of about 10 years, and are of different colours to show the different dates because they rust from the inside. In careless hands they could cause considerable personal injury and should be treated with respect. The recommended safe distance to stand clear of them is 30yd.

Wire Adjusters

Wire adjusters are used to adjust the tension on wires to signals that are a considerable distance from the signal box, as they expand and contract considerably at various times during the day. Because of the variations in temperature the slack may be needed to be taken out during the day and released in the evening. The adjusters are placed behind the

appropriate lever, and consist of a cast pedestal with a central screw thread which is turned by means of a handle on the top of the pedestal. They are usually painted black, with the lever number in white on the top. The screw thread has a slide on it, which goes down through the operating floor and raises or lowers

ABOVE A train describer. These were beautifully made instruments used in busy locations where any one of a number of routes may be sought. A despatch and receiving instrument would work in pairs. Few if any survive in use on heritage lines.

THE SIGNAL BOX

the wire according to the number of turns of the handle which are necessary to make the signal show the correct indication. Raising the slide reduces the slack, lowering it increases it. In cases of very hot weather a considerable number of turns may be needed to make the signal work correctly.

Lever Collars

Lever collars are reminder appliances which can vary in style according to the manufacturer of the frame. They are usually in the form of a cast round disc, painted red, about 4in in diameter. They have a hole in the middle of them so they can be slid over the top of a lever to make sure the signalman does not pull them by mistake, and also to remind him of the presence of a train or vehicles on the track.

ABOVE AND LEFT A signalman's lever collar. They are placed over the handle of a lever to remind the signalman not to pull the lever, for whatever reason, but in this case because there are wagons standing on the points. They come in various designs according to the company that manufactured them, but their most frequent use is as an egg cup!

ABOVE To compensate for expansion and contraction of signal wires in warm weather, wire adjusters are provided on certain signals. (Even the sun going behind a cloud for a few seconds can make a difference.) Seen here are WR and SR types.

THE SIGNAL BOX

THIS PAGE & OPPOSITE TOP
No two signal boxes will be exactly the same as regards their equipment, hence the need to 'learn the box' under the stewardship of an experienced trainer. In time the use of all instruments and levers will become second-nature. Note the difference in height of the levers; tall levers operate mechanical equipment, shortened levers are a reminder to the signalman that less effort is required as the item concerned is electrically worked.

LEFT Block bells and Block Switch. The Block Switch has a locking pin inserted which is to prevent accidental movement.

ABOVE Home and distant electrical repeaters showing both signals at ON. Notice the Distant Signal is lever No 0. This was an extra lever added to the end of the frame in Bewdley South when the Distant Signals were made operable. No 0 is the down distant and No 1 the associated down Home Signal.

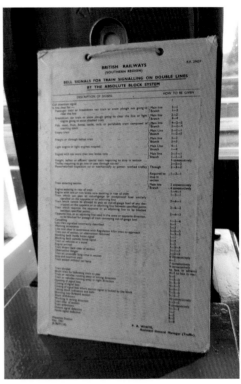

ABOVE Along with the associated Signal Box Footnotes the signal box should contain a list of standard bell codes – 'special' codes may be authorised at specific locations. The signalman should know from memory the codes that apply to his signal box AND, just as important, which bell code he may expect to receive next!

THE SIGNAL BOX

LEFT It has been mentioned that mechanical signals may be up to one mile from a signal box (some signals in BR days were in excess of this distance). There is, however, a limit to the distance points may be placed which in early days resulted in a signal box at either end of a station. Later, remote electrical operation was used, in this case with a 'hand-generator', or 'hurdy-gurdy' machine, as they were often known. The winding of the handle created an electric current to turn a point motor, which in turn moved the points and engaged the lock. An indicator at the top would show when this had been achieved.

RIGHT Old style first aid box; again part of the standard signal box equipment – the contents may well be modern.

RIGHT Detonator tin: sometimes detonators are kept secured to the brackets supporting the block shelf. Detonators are time sensitive, meaning they must be replaced at intervals, hence they are colour coded according to year of manufacture. If asked the question, 'What is the life of a detonator?' The answer is NOT 'One bang'!

ABOVE A 'King Lever'. This is used when a signal box wishes to switch out, and change from local to through working, enabling the signal boxes on either side to work with each other, or may be pulled to release a ground frame controlled by the signal box, in which it will lock certain levers in the frame.

LEFT Example of a Train Register sheet.

B.R. 24847/1

DOWN .... WESTFOLD MIDDLE .... Signal Box

.... TUES.day, ... 17TH day of .... JUNE 1963

Description of Train	REAR SECTION (Is Line Clear — Accepted under)									ADVANCE SECTION (Is Line Clear — Accepted under)							Remarks
																	I. Smith on duty 08:45
																	Detonators and Seals Released intact – Box in order
5-5~5																	Opened to East Box 09:01
16																	Sent and acknowledged to East Box 09:02 Instruments tested and in order
5-5-5																	Opened to West Box 09:05
16																	Sent and acknowledged to West Box 09:06 Instruments tested and in order
3-1 LOCAL	09:15			09:23	09:27					09:28				09:27	09:30		
1-4 GOODS	09:43			09:55	10:12									TO DOWN YARD			
4 EXPRESS	10:25			10:31	10:35					10:31				10:35	10:40		
1-4 GOODS				Ex DOWN YARD										10:45	10:47	10:55	
																	West Box advises driver of goods reports trespassers on the line by farm crossing
3-1 LOCAL	11:15			11:23	11:27					11:23				11:27	11:31		
																	Driver of local cautioned to look out for trespassers – Nothing found.
4 EXPRESS	12:09			12:15	12:19					12:15				12:19	12:24		
7-5-5																	Closed to East Box 12:20
7-5-5																	Closed to West Box 12:25
																	I. Smith off duty 12:30

Chapter 5

Rules and Regulations

Every railway is required to have a set of Rules, approved by the Government Railway Inspectorate, to which its staff must work. On heritage railways this is generally the one used by British Railways in its steam days, often with amendments to suit the particular railway concerned. There are about 240 rules altogether, which include everything from being told to report at the time and place appointed by the Company in a sober and orderly fashion to what must be done if a train fails on the line. But although an outline knowledge of all of them would be an advantage for anyone intending to work in a signal box, only a detailed knowledge of those applicable to signalling is necessary.

There will also be a book of Regulations and what form this takes will depend on what method of signalling is used on the railway concerned. If the railway only has one signalling procedure, then the book may only cover the one subject. If it uses two or more signalling methods the book will have to cover all methods. There may be one priority set of Regulations, for example Token Working on single line, which must be learned first, before a signalman is allowed to progress to another such as Absolute Block Working for double line. However, although there will be variations in the methods of working, the basic theme will still be the same.

After the Book of Rules and the Regulations, both of which cover the whole railway in broad terms, there is the Sectional or General Appendix. This itemises specific conditions and locations where Special Instructions apply, such as speed restrictions and locomotive route restrictions.

Finally, there are the local instructions, which are modifications or amendments to the Rules and Regulations, applicable to the particular signal box concerned. These are known as the Footnotes and are generally clipped or pinned to a board in a prominent place on a wall within the box.

All of this has to be learned by the signalman, and he will be subject to an oral, or possibly written, examination on these subjects before being allowed to take duty on his own. Most of it is very boring to read, and will seem like common sense, but a casual knowledge of them all is useful in understanding the problems faced by all those involved in running the railway. The suggested method of learning is to spend approximately 30 minutes each day studying a particular section of a book as a more concentrated reading can leave the head spinning, and the mind very muddled!

When the signalman has reached an acceptable level of knowledge of the Rules and Regulations a re-examination can occur at any time, but usually at regular intervals such as every two or three years, in order to comply with Railway Inspectorate requirements.

The Rule Book

The Rule Book can be in many styles, according to the wish of the operating company. The British Railways book was about 5in wide and 6in deep with a black cover, and the rules were divided into sections relevant to the occupation of those working on the railway. The first section was general, applicable to all employees, advising them what they could and could not do as well as including terms of employment, medical requirements and disciplinary procedures.

The next section dealt with the working of stations and was meant for Station Masters, who had men under their control and were responsible for the upkeep and condition of the buildings and platforms as well as the movement of trains.

After that came the section that applied to signals and signalling, and it is this section that someone who wishes to become a signalman needs to study. It deals with fixed signals and the indications they give; hand signals, what they mean and where they should be given; what to do if trains are detained on running lines; use of detonators; how points and signals should be worked; the procedure for working if equipment is disconnected or defective; working in fog or falling snow; and working at station yards and level crossings – in all about 75 rules.

Other sections deal with shunting; head and tail lamps; working of trains, which includes the duties of the driver, fireman and guard; reporting of accidents; trains stopped by accident or failure; and finally, permanent way and works.

The rules are very specific in outlining what can and cannot be done, but at the same time cannot cover every situation so it is up to the person dealing with the circumstances at the time to interpret them in the spirit in which they have been written.

Regulations for Train Signalling and Signalmen's General Instructions

The Book of Regulations will again vary according to how the company wishes the Regulations to appear, and will contain details of all the methods of signalling used on the railway to which they apply. If the line is single the first part of the book will contain instructions on how to work the single-line equipment, and an outline of the method of working for signalling the trains, together with a list of bell-codes to be used and how entries are to be made in the train register book. Then comes a long list of Regulations, similar to those in the Rule Book but more comprehensive, setting out the correct procedure for signalling and what to do in special circumstances and emergency situations. Depending on what methods of signalling are being used, there will be approximately 32 Regulations. If more than one signalling method is being used, the book will follow the section on the main method used on the railway with another section on the other methods used. There will then be a section at the end of the book called Signalmen's General Instructions, stating how the various items of equipment in the box should be operated and in what circumstances. It will also show examples of the forms to be used in the event of emergency working or failure of the equipment.

Having dealt with how to use the equipment, the method of working and any special bell-codes, the Regulations then go on to describe the conditions under which a train can be accepted from the box in the rear, and then deal with progressively more complicated situations. They describe what should be done if it is necessary to institute Fog Working and under what circumstances it should be brought in to operation, how to deal with shunting procedures, engineers trains, and trolleys on the line, then how to deal with exceptional circumstances such as an obstruction or animals on the line, or a train failed in section. There are Regulations for trains passing without tail lamps, examining the line, trains dividing or breaking and running away. The book also covers failure of equipment, the forms needed for such eventualities and how to deal with such situations as pilotman working, and, if authorised, shunting into forward section and wrong direction working.

Finally, at the back of the book, are Signalman's General Instructions, which detail how ancillary items such as lever collars and reminder appliances should be used, as well as other covering other topics such as telephones, working in fog and falling snow, recommendations how to work the equipment in times of severe weather, the general housekeeping of the box and keeping everything in order.

Sectional, or General, Appendix

The Appendix contains all the variations necessary to the standard Rules and Regulations, which need to be applied to any local situations. It is a considerable document, giving instructions for everything involved in the running of a railway. Details of signalling configurations for specific circumstances are included, as are requirements for any shunting moves which

RIGHT, OPPOSITE PAGE AND OVERLEAF As recounted earlier, Ground Disc Signals sometimes can be placed one on top of the other, meaning they apply to differing routes on the basis of 'top to bottom left to right'. For many years this was also the way of describing signals at a junction; hence the two stop arms refer to different routes and may thus appear in any of the three combinations seen. Note that on a tall post where sighting might otherwise be difficult over a distance, two stop arms that act together may appear on the same post. The driver thus has an advance warning of the signal as he approaches and when he is close by he may use the same signal nearer to eye level. (A modern day alternative is the banner repeater.)

may have to be made, involving occupation of the main running lines. It also covers the use of the vacuum brake, and air brake if necessary, how to exchange Tokens, use of the Token exchanging apparatus where installed and greater detail on the working of points and signals and the care and maintenance of signal boxes. There are sections on the working of trains (both passenger and freight), the tools required on board, the formation of trains, conveyance of merchandise and how certain loads must be carried. Attaching and detaching of vehicles, permanent way works and equipment, what has to be done in the case of an accident, lineside fires, fog and falling snow, and station working are also included, as are items such as speed restrictions over sidings, whether some types of locomotives are permitted to enter those sidings and the capacity of any loops or sidings in terms of the numbers of wagons or coaches. There may also be a route details section showing speed restrictions and locomotive restrictions along the line, as well as the capacity of any running loops at stations where trains are allowed to cross.

Signal Box Footnotes

In addition to all of the previous Rules and Regulations, there is always a set of Footnotes which apply to each particular signal box because

every box is different, with its own geographical location. These cover the specific methods of working either side of that box, any special bell-codes that are used, whether any Rules and Regulations that can only be used if authorised are applicable, fog marking points, working of signals at that particular box, conditions to be applied in fog and falling snow and any special restrictions that may apply to sidings and goods yards worked from that box.

Unwritten Rules

As well those Rules and Regulations set out in the various books there are some which are 'unwritten' and which promote good practice when dealing with trains and movements of rolling stock on a railway.

Perhaps the most important of these is that it is considered very sensible never to wear any bright red or green clothing when working anywhere on a railway, whether dealing with moving trains or not, in case a driver takes the clothing as an emergency flag.

Also, when walking alongside stabled vehicles and wishing to cross the track immediately behind them, never do it alongside the buffers. Always allow some space between the stock and yourself, in case an engine has been put on the other end without your knowledge and then pushes the vehicles back.

Furthermore, when walking along the railway line always keep your eyes and ears open. If possible, always walk facing the direction from which the traffic will be coming so that you can see any train coming towards you. In this event, when a driver sees you he should give you a short whistle to tell you he is coming, which you should acknowledge by raising your arm above your head. This tells him that you have seen him and will stand clear; he should then give you another short whistle to say that he has seen your hand signal. If you are working on a platform and a train approaches, you should follow a similar procedure. It is then good policy to take a couple of steps back and stand at least 4ft from the platform edge to avoid any turbulence from the train.

All of this may seem like a lot to understand and, if not tackled correctly, can deter all but the most ardent trainee, so it may be easiest to learn a small section at a time, ideally combined with some practical experience. Then it will be realised that most of it is common sense and good housekeeping, and once the basic elements of signalling and safety procedures are understood, the remainder comes as a matter of course.

ABOVE A bracket signal made from concrete with secondary 3ft arm attached for a lesser route.

Chapter 6

Local Signal Box Conditions

As has been mentioned earlier, no two signal boxes are the same. The style and method of construction may appear the same but each one is in a different geographical location and has a different track layout, and so will have different operating amendments peculiar to that signal box. These amendments are known as the 'Box Footnotes', or 'Special Instructions', and are usually posted to a suitable board, fixed to a blank wall within the box. Some railways are now enclosing them in a loose-leaf binder, kept in the desk for safe keeping.

Footnotes

These instructions will detail the methods of signalling on either side of the box, and any special bell-codes that may be used, such as adding a 1pause3 to the standard code, for trains requiring specific routes, at a box in advance. They will also state to which point the line must be clear before a train can be accepted from the box in the rear. If the box is a crossing place on a single line, the line will usually need to be clear to the

Starting Signal on the loop line. But if the box is also signalled for through running, and a through train is being signalled, then the point will be 440yd in advance of the outermost Home Signal. This point is known as the standard Clearing Point, and always requires that a train has not been accepted in the opposite direction. In certain circumstances, there may also be permission to accept a train when the line is not clear for 440yd ahead of the outermost Home Signal, but in these cases, known as accepting under the Warning Arrangement, the driver must be advised of the situation by the signalman in the rear.

Fog Working

Every signal box has what is called a 'Fog Marking Point', which is generally a signal about 200yd from the box, which the signalman must be able to see for normal working. If, for reasons of poor visibility, he cannot see the specified signal then 'Fog Working' may have to be introduced. This entails posting fog signalmen at the Home Signals, with detonators to place on the line, to prevent the approach of a train. Fog Working also requires much more restrictive working, such as bringing both trains to a stand, before allowing one in to the station, if trains are to be crossed. If trains are not crossing, and a through train is offered to the signal box, then, unless otherwise specified, train out of section for the previous train must have been received from the box in advance before the train can be accepted from the box in the rear. Precise details for Fog Working can vary from railway to railway and even signal box to signal box, so a study of the Rules as applicable to each individual signal box should be made in order to ascertain correct procedure.

Occupation of the Running Line for Shunting Purposes

The Footnotes will also give any instructions necessary for the shunting of sidings at a signal box, particularly if it means occupying the running line, where the driver may need to be in possession of a Token for such purposes and precautions need to be taken to avoid delay to any service trains. In these circumstances knowledge of the vehicle capacity of any sidings, in terms of the number of wagons that can be stored on them, can be very useful to avoid unnecessary moves and delay to trains. It is also useful to know the position of the loading gauge, and any locations in the goods yard where stabling of vehicles would be foul of any other sidings.

Wrong Direction Running

In most cases, all trains are run in the direction for which they are signalled, known as 'Right Road Running', but the Footnotes may, in special circumstances, allow wrong direction running. This requires special bell-codes and possibly co-operation from the signalmen on either side. Wrong direction running is usually done for convenience of operation, simplicity of moves and to save time, and specific written instructions are usually laid down.

Failure of Equipment

Although the Book of Regulations lays down the procedures for dealing with failures of equipment, such as instituting Pilotman working, when the Token equipment fails the Footnotes may expand these details to take account of any special local situations that may exist at a particular signal box, such as whether or not any track-circuits show clear and whether or not the train has been correctly accepted by the box in advance. The most important thing in the case of equipment failure is to inform the Duty Officer, or whoever is in charge of the railway that day, so that alternative arrangements can be made. The Signal and Telegraph Department should also be advised so that the equipment can be put in order as soon as possible.

LOCAL SIGNAL BOX CONDITIONS

RIGHT This is a closer look at a guide for point rodding, as seen in the left hand illustration on page 73. The concrete 'I' section is known as a stool and is sunk in the ground until its head is just above the surface. The metal guide is known as an 'A' stand, and the rodding runs within the stand on the rollers, and they must not NOT be oiled, to prevent accumulation of dirt and grease. To the casual observer it may look simple to install, but as with most things it is more difficult than it seems at first.

Switching a Signal Box In and Out

Normally, a signal box is opened at the beginning of the service and closed at the end of it, but if a signal box can be switched out and through working introduced between the boxes on either side then the switching in and out procedure will be laid down in the Footnotes and the method of obtaining the co-operation of the signalmen on either side will be specified. This is known as going from long to short section working, or vice versa. In these cases, switching a signal box in and out of circuit is usually only done when the timetable demands it, to enable trains to be crossed. There may also be Special Instructions for access to any sidings, either in the section or at the signal box, when it is switched out.

General

If there are any local circumstances that need Special Instructions, these will also be included in the Footnotes. They may cover such things as the working of any barrow crossings or minor level crossings in the sections either side of the box where, if the crossing is manned, the crossing keeper needs to be advised of the approach of a train; any peculiarities concerning exchanging of Tokens, such as whether the Tokencatchers are needed, which may be the case with diesel-hauled trains, and whether Token exchanging can be done by two people, one to collect the Token being given up, the other to hand over the Token to be delivered. The signalman may also have to take special precautions, such as advising station staff, or train crews, when men are working either below the box or out on the line.

There will also be details of any locomotive restrictions, where they can and cannot go and whether the signal boxes either side need to be advised in the case of double-heading of trains or trains being assisted in rear. It is also worth learning the various locomotive headlamp codes, certainly the most frequently used ones, and to bear in mind that they may be changed on route if the classification of a train changes. For example, a train may start its journey as empty coaching stock until it reaches a station where it is required to pick up passengers, then it will continue its journey as a passenger train, meaning a different headlamp code.

Similarly to locomotive restrictions, there will be notification of any permanent speed restrictions on the track and whether they apply just to certain locomotives.

In working a signal box, knowledge of the section times either side of the box will come as a matter of course, as will the 'feel' of each lever when being pulled. The working Fouling Bars may be heavy at first,

but once pulled over half-way the remainder of the stroke will fall as the bar drops down and takes the lever with it. It is also helpful to have knowledge of the workings of the boxes on either side because much of the unseen art of signalling is anticipating two or three moves ahead in order to keep the trains moving. A copy of the Working Timetable, together with any Special Traffic Notices, can be a very useful aid in achieving this objective.

Most signalman are required to work a signal box for which they are qualified a minimum number of times per year, in case the local instructions are amended, but if the working is very infrequent it is a good idea to make a copy of all the relevant Footnotes and Special Instructions so that the signalman can refer to them when off-site.

It is also not unknown for signalmen to keep with them an Ordnance Survey map, or a Line of Route Guide, so that in an emergency staff can be directed to the specific location by map reference, which saves much time and eliminates any confusion.

Good housekeeping is also part of being a signalman and such things as keeping the box tidy and free from dust and dirt all play their part, as is using a lever cloth to keep the handles shiny to avoid incurring blisters on the hands and cleaning windows so that signals and points can be easily seen, particularly at night or if the sun is in the wrong position during the day.

ABOVE AND FACING PAGE Point clips. As referred to in the text, if it is necessary for a passenger train (not empty coaching stock – meaning passenger coaches without passengers) to make a facing move over a point which is not fitted with a lock OR if the lock is defective, then before this is permitted the facing point MUST be 'clipped and padlocked'. A more accurate term might be 'clamped and padlocked', for a point 'clip' is placed under the running rail and then screwed tight so as to prevent the point blade from moving. A padlock is then applied to prevent further movement. Note there are different types of point clip dependent upon whether bull-head or flat-bottom rail is in use. Any such move would be undertaken at low speed and under the direction of a man on the ground with whom 'a clear understanding had been reached' involving the man on the ground, driver and signalman. The working of a passenger train over a facing point in this way would be unusual to say the least, but may be necessary in the case of failure of the equipment.

Chapter 7

Emergency Situations

Throughout any railway safety is absolutely paramount and over-rides all other considerations, especially where the running of trains is concerned. With current Health & Safety legislation also making itself felt, safety and personal welfare is also a high priority in all other departments as well, from such major items as engineering and maintenance equipment right down to safe walking routes to and from outbuildings such as signal boxes.

With regard to train movements, should an emergency arise it is good policy to remember that delay is better than derailment. If a signalman is unsure which procedure he should follow, best practice is to stop the trains and read the Rule Books because if the trains cannot move they cannot cause any harm. Then careful assessment of the situation can be made and correct decisions taken to deal with the circumstances.

Obstructions

The most serious danger to traffic is an obstruction on the line, and if a signalman is advised of such a thing the correct bell signal to send is six beats consecutively to the appropriate signal box, or to the boxes on either side if necessary. The sending signalman should then advise the man to whom it was sent the reason for sending the signal. The same procedure should also be adopted if the signalman becomes aware of a track defect as this also represents an obstruction to traffic. The signalman receiving the bell signal should take all necessary measures to prevent trains from proceeding towards the box from which the emergency signal was received, such as replacing his signals to danger and placing detonators on the track if required. He should then answer the bell signal, the reasoning being that in such situations the stopping of trains is more important than answering bell signals. Having stopped the traffic the relevant departments should be informed, such as Duty Officers or whoever is in charge of the railway on the day as well as the Permanent Way staff, in order that the obstruction can be removed, alternative arrangements made for any passengers and normal working resumed as soon as possible.

Examination of the Line

Having become aware of an obstruction on the line and successfully stopped all the trains, the line now has to be examined. This is where the local conditions at the signal box are taken into consideration. It is easiest to examine the line with an engine, because locomotives can be stopped abruptly if anything is found, and if possible a responsible person, such as a guard or Station Master, should accompany the examining party. If information has been received that a passenger has fallen from a train, the engine should always be so accompanied and should have a brake van attached to it, of either the goods or passenger type.

In some circumstances it is possible to use a passenger train or a train with fully-fitted brakes on all vehicles to examine the line, but this may be precluded if there is a tunnel in the section or visibility is reduced on the portion of line to be examined.

The sequence of events for carrying out this procedure is first of all to advise the signalman in advance of what is about to be done, usually by means of the internal telephone, but if this is not available any other type of telephone including BT and mobiles can be used. If Token Working is in operation it is necessary to obtain a Token from the signalman for the section of line to be examined. It is not a requirement to ask for a Line Clear, in the normal way, because it has been established that the line is not clear. Having obtained a Token, the driver of the locomotive or train examining the line should be advised of the situation, and be told to proceed with caution. He should be prepared to stop short of any obstruction, and where the obstruction is likely to be, if possible by reference to the mileposts or signals. The signalman should then send the train entering section bell signal – two beats – to the box in advance. Once the signal has been acknowledged the driver can be told to pass the Starting Signal at danger and go ahead and examine the line. The driver must examine the whole section, but having done so can either proceed to the box ahead or return to the box in the rear, whichever is most expedient. The important thing to remember is that the Token must not be placed in the Token instrument unless the line is reported as being clear in order to prevent another train being signalled in to the section.

If the line is reported as being clear the next train through the section can be signalled normally but it would be wise to advise the driver of that train of what has happened and ask him to keep a good lookout.

If the driver does not report the line as being clear, what happens next is beyond the control of the signalman as it depends upon the specific circumstances and how they are to be dealt with. The signalman can only report the situation and the findings of the driver to those in charge of the railway and await instructions.

EMERGENCY SITUATIONS

Disabled Train

If a train fails, or becomes disabled in a section, how it is dealt with will depend on the Regulations of the railway concerned. For many years, if the failure occurred on a single line the Staff, Key or Token had to be taken to wherever the assistance was available but in recent times this has changed. The Regulations were changed on the national network because some sections can be up to 20 miles long and communication via mobile phones and radios is much improved so the Token now stays with the disabled train. Some heritage railways have adopted these changes so reference to the appropriate book of Rules and Regulations should be made before taking any action.

When a train or portion of a train fails in section the first thing to establish is where the failure has occurred and confirm that detonators have been placed on the rails to protect the stationary train. This is usually achieved by a member of the train crew walking to the box in the rear or the box in advance (whichever is the nearer) and informing the signalman of what has occurred. In some cases, even if the failure occurs in advance of the outermost Home Signal, the local box conditions may still require the Regulations to be applied as if the train

RIGHT Carriage door handle not turned/closed properly. The signalman MUST observe the train as it passes, watching for any signs of alarm or incident – such as a door handle not turned or closed properly. (A number of preserved lines have heritage stock on which the door handle must be turned to lock the door.) If such an incident is observed the train should be stopped if it is safe to do so – but not for example on a viaduct – or the 'stop and examine' signal sent to the signal box in advance.

had failed in the section and not within fixed signals.

Having established the location of the failure, the Duty Officer or Station Master should be informed as soon as possible to establish from where the assistance will be obtained. This information can then be passed to the train crew of the failed train and to the signalman at the other end of the section.

Presuming the Regulations say the Token must accompany the assisting train, and it has been decided that the assistance will come from the rear, the fireman of the disabled train will have walked back to the rear box,

checked that the protection is in place on the way and shown the signalman the Token. The signalman will then advise the signalman in advance of what is about to happen, send the train entering section bell signal to the signalman ahead and, once it has been acknowledged, give the Token to the driver of the assisting train. The driver will be told to pass the Section Signal at danger and the fireman of the disabled train can ride on the engine of the assisting train to conduct its driver to where the train has failed. If the assisting train propels the disabled one to the box in advance, the signalman there can send the train out of section bell signal (2pause1) to the box in the rear, provided both trains have arrived complete. The next train through the section can then be signalled in the normal way but the train must be stopped and the driver advised of the circumstances and instructed to proceed with caution. If the disabled train is drawn back to the box in the rear, the cancelling bell signal (3pause5) must be sent to the box in advance because the train has been withdrawn

from the rear of the section. Again, the next train through the section must be stopped and cautioned.

If, however, the assisting train has to come from the box in advance, then the signalmen must confer, and the Token must be placed in the instrument by the signalman in the rear, in order for the signalman ahead to be able to withdraw one at his end to give to the driver of the assisting train, going in from the advance end of the section. In these cases the driver of the assisting train must be told where the failure exists, and be prepared to meet either the fireman or guard, who will then accompany him to the disabled train.

As with the failure being withdrawn to the rear, if the failure is then taken forward the next train through the section can be signalled normally but the driver must be instructed to proceed cautiously.

Should the failure, through force of circumstances, be divided and a portion be withdrawn to the rear whilst the remainder was taken forward, the section of line has not then been proved clear of all obstructions so it will be necessary to carry out an examination of the line, as detailed previously. Once a train has passed through the section, complete with a tail lamp, normal working can be resumed.

If the railway's Regulations have been amended to say that the Token must stay with the failed train, study of the Regulations will be necessary because the driver of any assisting train may have to be handed a form, or some paper work, which will instruct him to proceed to an occupied section and tell him the location of the failure as well as to which end of the section it should be taken.

Animals on the Line

It is a legal requirement for all railways in this country to be fenced but nevertheless fences can break down, giving an open invitation to animals to trespass on the railway. In the event of a report of animals on the line, the first thing that must be done is to inform the signalman at the other end and take steps to have the section cleared by any available members of staff and the fence repaired by the Permanent Way department as soon as possible. In the meantime, all trains should be brought to a stand and the drivers instructed to proceed cautiously. If there is a tunnel in the section drivers should not to proceed in to the tunnel until they have established that it is clear.

Train Unusually Long Time in Section

If a train is in a section for a significantly longer time than normal the signalmen should confer and if necessary inform the Station Master or Duty Officer. All traffic should be brought to a stand until word is received from the train crew of any problems. The traffic should remain at a stand until the reason for the delay is known and it can be decided what action should be taken.

Observation of Trains

It is a requirement of all signalmen to observe the passage of trains as they go past the signal box, in case

ABOVE It is a requirement of the signalman to observe the tail lamp of every train, because that is the only indication he has that the train is complete. Should a train pass without one, this indicates that the train has become divided, then the bell signals are 9 to the box in advance, where the train must be stopped and examined, and 4-5 to the box in the rear. These pictures taken on the Swanage Railway show a train without a lamp, and the guard placing one in position on the bracket before the train departs.

there is any need to stop the train at the box ahead and also to check that there is a tail lamp on the last vehicle, which indicates that the train is complete. The most common cause for stopping a passenger train is a door open, or just left on the latch and not properly closed. In these circumstances the guard has to establish whether a passenger has fallen from the train, and if so it becomes necessary to carry out an examination of the line. If a train passes without a tail lamp the signalman has to advise both the man in the box ahead, by sending the 'Train Passed without Tail Lamp' bell signal, 9 beats consecutively, and the man in the box in the rear, for which the bell signal is 4pause5. The man ahead, having stopped the train, should then ascertain from the guard

EMERGENCY SITUATIONS

ABOVE LEFT A 'tell-tale'. Normally horizontal, this indicator will move to show on which coach the emergency alarm chain has been pulled.
ABOVE CENTRE Observation: here the key supporting the track in the rail chair has become displaced. The permanent-way department should be informed. **ABOVE RIGHT** A detonator affixed to the track. **OPPOSITE** Modern day STOP board advising of a temporary obstruction.

whether the train is complete. If not, he should advise the man in the rear, who sent the 9 bells signal, because the train may have become divided and steps will have to be taken to deal with the rear portion, such as diverting it in to a siding to prevent it coming in to collision with the front portion. No train should be allowed to proceed towards the divided train until the circumstances have become clear and it is considered safe to do so, which

may mean having to rescue a portion of a train or at the very least arranging for an examination of the line.

Divided Trains

A loose-coupled freight train, which only has brakes on the engine and guards brake van, will carry a red tail lamp, and a red side lamp on either side of the rear brake van. If any of these are missing as the train passes

is on a falling gradient it may run through the section to the signal box ahead.

Runaway Trains

In extreme cases, particularly with loose-coupled freight trains, there is a possibility of the locomotive not being able to keep the train under control on a down gradient. This is usually heralded by the driver giving a long continuous blast on the whistle as he approaches a signal box, and then the signalman has to make some quick decisions on how he should deal with the situation. If possible, it is best to obtain permission from the box ahead to allow the train to run, if time allows, by asking for a 'Line Clear' and obtaining a Token so the driver can proceed in to the section and bring his train safely to a stand. Should that not be possible, or if, on a single line, the signalman has accepted a train in the opposite direction, the runaway train will have to be dealt with as expeditiously as possible in order to prevent any collision, and minimise any damage to locomotives and rolling stock. There are two bell-codes for runaway trains: 2pause5pause5, which is used in either direction on single line, and 4pause5pause5, which is used in addition on double track. The 2-5-5 is used on double track for a train running away in the wrong direction, and the 4-5-5 for a train running away in the right direction. The most convenient way to remember them is to think of 2-5-5 being the shorter message to send, in a greater emergency, which is a train running away in a wrong direction.

a signal box it must be presumed at once that the train has become divided and steps should be taken immediately to prevent the two portions coming in to contact with one another. If this cannot be done in sufficient time the train crew must be advised, usually by a signalman waving a green flag from side to side, so that the speed of the front portion can be regulated to minimise the effects of the collision with the rear. It is in these situations that knowledge of the line of route and gradients becomes very useful because if the breakaway portion is running on a rising gradient it may come to a stand in the section, but conversely if it

In cases where a train enters a section without a Token or passes a Starting Signal at Danger the 'Train Running Away' should still be sent because the driver does not have the authority to proceed, but it may not be necessary to carry out any emergency measures to divert the train. Such cases are generally very rare and it usually results in severe disciplinary action being taken against the train crew.

Failure of the Token Equipment

In the event of failure of any equipment immediate steps should be taken to have it repaired, but in the case of the Token instruments breaking down special measures have to be introduced to enable the passage of trains to continue. This means introducing working by Pilotman, and advising all concerned that it is in operation. A Pilotman is, in effect, a human Token, and no train can proceed in to the section without him being present or him giving the driver written authority to go in to the section ahead. Whoever puts Pilotman working in to operation has to provide a written form to the Station Master at each end of the section, the signalmen at each end of the section and the Pilotman. The signalmen and Station Masters at each end of the section must sign the Pilotman's form to acknowledge that such a method of working is in operation. The precise details of how the procedure is to be put into operation are quite strict, and reference to the Railway's Book of Regulations should be made in order to ensure

OPPOSITE Train starting away against the Stop Signal – or has the signalman replaced the signal far too early? If the former the 'Train Running Away on Right Line' bell code would immediately be sent to the box ahead although the signalman would also attempt to attract the attention of the engine crew and guard. (The guard has his own responsibility to observe the position of signals.) Note – this was a deliberately staged incident under full control.)

ABOVE Track circuit clips are heavy duty metal clips joined by a length of wire, and clipped over the head of the rail in an emergency, on sections of line which are track circuited. This attracts the attention of the signalman, and protects the rear of the train by preventing the signals behind it being put to clear.

that everything is done correctly and that there is no misunderstanding on anyone's part about what is happening. If there is a succession of trains to go through the section in the same direction the Pilotman can send them through, one at a time, on a Pilotman's ticket. The signalman needs to have planned three or four moves ahead before this happens to make sure the Pilotman is not at the wrong end of the section from his signal box when a train needs to be despatched and a consequent delay is incurred.

A Selection of the more Common Bell Codes

Call attention   **1**

'Is line clear?' for:

Class

1 Express passenger train, postal train, newspaper train, or breakdown van train
or snow plough going to clear the line, or light locomotive going to assist disabled train   **4**
Officers' Special train not requiring to stop in section

2 Ordinary passenger train, mixed train, or breakdown van train or snow plough NOT going to clear the line   **3-1**

5 Empty coaching stock train (not specially authorised to carry Class 1 headcode)   **2-2-1**

9 Unfitted freight train (where authorised)   **1-4**
Freight train, Officers' Special train or Engineer's train requiring to stop in section   **2-2-3**

0 Light locomotive, light locomotives coupled or locomotive with brake tender(s)   **2-3**

Train entering section   **2**

Train approaching (where authorised)   **1-2-1**

Cancelling   **3-5**

Last train signalled incorrectly described   **5-3**

Warning Acceptance   **3-5-5**

A SELECTION OF THE MORE COMMON BELL CODES

Line now clear in accordance with Regulation 4 for train to approach	**3-3-5**
Train out of section, or Obstruction Removed	**2-1**
Blocking back inside home signal	**2-4**
Blocking back outside home signal	**3-3**
Train or vehicles at a stand	**3-3-4**
Locomotive assisting in rear of train	**2-2**
Locomotive arrived	**2-1-3**
Train drawn back clear of section	**3-2-3**
Obstruction Danger	**6**
Train an unusually long time in section	**6-2**
Stop and examine train	**7**
Train passed without tail lamp to box in advance	**9**
to box in rear	**4-5**
Train divided	**5-5**
Shunt train for following train to pass	**1-5-5**
Train or vehicles running away in wrong direction	**2-5-5**
Train or vehicles running away in right direction	**4-5-5**
Opening of signal box	**5-5-5**
Closing of signal box	**7-5-5**
Closing of signal box where section signal is locked by the block	**5-5-7**
Testing block indicators and bells	**16**
Shunting into forward section	**3-3-2**
Shunt withdrawn	**8**
Working in wrong direction	**2-3-3**

The above are the most common Bell Codes used today, but some Heritage Railways use Big Four (before 1948), or even Pre-Grouping (before 1923) codes, so in practice there may be differences from those listed above.

Questions for Signalmen

Chapter One

1). What is the maximum age, before special permission is required to work with moving trains?

2). What is meant by 'Learning a Box' and 'Learning a Frame?'

3). How long should a newly qualified signalman work in a signal box before considering moving to another one?

Chapter Two

4). What is meant by a 'Block' in 'Block Working'?

5). What is a 'Clearing Point'?

6). On a single-track railway, what is the driver's authority to be on the line?

Chapter Three

7). What is meant by a signal being 'Off'?

8). Name three types of fixed signals.

9). How is a Distant arm different to a Stop arm?

Chapter Four

10). What is the purpose of a 'Signal Box Diagram'?

11). What does a red lever do in a frame?

12). What does a blue lever do in a frame?

Chapter Five

13). What is contained in the Sectional or General Appendix?

14). What are Signal Box Footnotes?

Chapter Six

15). What is a 'Fog Marking Point'?

16). What is meant by 'Wrong Direction Running'?

17). When would working by Pilotman be introduced?

Chapter Seven

18). What is the most serious danger to traffic on the line?

19). What is not required when a train has to examine the line?

20). What is the first requirement if a train fails in section?

Answers...

Chapter One

1. 70 years of age
2. To learn the workings of a particular signal box includes not just the actual Lever Frame, position of levers relative to signals, points etc, but also operational requirements unique to that signal box. Learning a frame means getting to know how to pull each lever, and what each lever does.
3. A minimum of one year. This period may be extended dependent upon how many turns have been achieved or at the behest of the signalling inspector.

Chapter Two

4. The section of line beyond the most advanced Starting Signal of one signal box and the outermost Home Signal of the next signal box.
5. A portion of line, usually 440yd, ahead of the outermost Home Signal, which must be clear before a train can be accepted from the box in the rear.
6. The Token (includes Staff/Tablet) or the presence of the authorised Pilotman on the engine. (Does not apply when shunting within Station Limits.)

Chapter Three

7. The signal has been cleared from the stop or caution aspect. It means it is displaying a proceed indication.
8. Stop Signal, Distant Signal, Ground signal.
9. A distant arm is yellow in colour and has a fishtail end with a similar shaped black band, whereas a stop arm is red and end with a square end and white band.

Chapter Four

10. A diagram of the track and signal positions applicable to the area controlled by the signal box and the levers therein, and, if required, to indicate the position of trains.
11. Works a Stop Signal.
12. Operates a lock or Fouling Bar.

Chapter Five

13. The amendments to the Rule Book to be followed for the safe working of a particular location and/or section of line.
14. Local instructions to be followed concerning the operation of a particular signal box or Ground Frame.

Chapter Six

15. A fixed object, usually 200 yards away, which under normal conditions may be easily visible from a signal box. If this set object is no longer visible due to adverse weather then fog-working, as applicable, must be introduced.
16. A train operating in the opposite direction to what may normally be expected, i.e, moving 'up' on the 'down' line.
17. The failure of Token ('Tablet/Staff') instruments or when such equipment is temporarily out of use.

Chapter Seven

18. An obstruction.
19. A 'Line Clear'.
20. To establish the position of the failure, and that it has been protected, then to advise the signalman at the opposite end of the section concerned and arrive at a CLEAR UNDERSTANDING as to the action to be taken.

Index

Photographic credits:

All photographs are from the author's collection unless credited otherwise.

Design & Artwork: ALEX YOUNG

Published by: DEMAND MEDIA LIMITED

Publisher: JASON FENWICK

Written by: DAVE WALDEN